Spiking Neural P Systems for Time Series Analysis

Spiking Neural P Systems for Time Series Analysis

Jun Wang
Xihua University

Hong Peng
Xihua University

IEEE Press
WILEY

Published by John Wiley & Sons, Inc., Hoboken, New Jersey.

Library of Congress Cataloging-in-Publication Data has been applied for:

Print ISBN: 9781394381579
ePDF ISBN: 9781394381593
ePub ISBN: 9781394381586
oBook ISBN: 9781394381609

Cover Design: Wiley
Cover Image: © Dmitry/stock.adobe.com

Contents

About the Authors

Jun Wang is a professor at the School of Electrical Engineering and Electronic Information at Xihua University, Chengdu, China. She is a member of the "International Membrane Computing Society (IMCS)." She is the main investigator of 3 scientific research projects funded by the National Natural Science Foundation of China and of more than 30 scientific research projects at the national and provincial levels. She was awarded the Sichuan Provincial Natural Science Award in 2017. Her research interests cover several topics, including membrane computing, time series analysis, artificial intelligence, machine learning, intelligent control, and power systems. She has published over 90 scientific papers in international journals and conferences. She has more than 3,000 citations with an H-index of 35, according to Google Scholar.

Hong Peng is a professor at the School of Computer and Software Engineering at Xihua University, Chengdu, China. He is a member of the International Membrane Computing Society (IMCS), IEEE, and CCF. He is the main investigator of 4 scientific research projects funded by the National Natural Science Foundation of China and of more than 20 scientific research projects at the national and provincial levels. He was awarded the Sichuan Provincial Natural Science Award in 2017. His research interests include membrane computing, artificial neural networks, time series analysis, deep learning, image and computer vision, and natural language processing. He has published over 230 scientific papers in international journals and conferences. He has more than 4,900 citations with an H-index of 42, according to Google Scholar.

Preface

Time series analysis has become an invisible pillar of modern scientific and engineering research. Essentially, it processes a set of data points arranged in chronological order, aimed at extracting valuable information, identifying potential patterns, and predicting future trends. In the field of meteorology, it is through time series modeling of parameters such as atmospheric pressure, temperature, and wind speed that we can achieve relatively accurate weather forecasts. In financial engineering, the time series analysis of stock prices, trading volumes, and market indices forms the basis of quantitative trading. In the field of neuroscience, functional magnetic resonance imaging (fMRI) maps the functional connectivity of different regions of the brain by analyzing the temporal correlation of blood oxygen level dependent (BOLD) signals. Neuroscientists are attempting to uncover warning signals for epileptic seizures through millisecond level electroencephalography (EEG) fluctuations recorded by electroencephalography.

With the explosive growth of the Internet of Things (IoT) and big data technology, time series analysis is facing unprecedented opportunities and challenges. Modern industrial equipment may be equipped with hundreds of sensors, each generating thousands of data points per second. The urban transportation system generates billions of vehicle location time series records every day. Dealing with this type of "time series big data" requires the development of new algorithms and computing architectures.

Membrane computing, as a branch of natural computing, has become a promising paradigm for designing non-traditional computational models. Within this framework, spiking neural P systems (SNP systems, in short) stand out due to their biologically inspired architecture, parallel processing capabilities, and dynamic spiking mechanism. These systems are rooted in the functions of brain neurons, providing a unique approach for modeling and solving complex computational problems. Due to their inherent parallelism, dynamic spiking behavior, and ability to model discrete events in a biologically plausible manner, the SNP systems have received attention. With their dynamic modeling capabilities, they are particularly suitable for processing time series and other temporal data.

The book, *Spiking Neural P Systems for Time Series Analysis*, explores the intersection of these two fields and proposes a new method for processing, analyzing, and predicting time series data using the SNP systems. Our goal is to bridge the gap between membrane computing theory and real-world time series problems by comprehensively discussing the recurrent-like and echo-like models of the SNP systems, as well as case studies, to demonstrate their effectiveness in time series analysis. Case studies include financial time series analysis, power load forecasting and photovoltaic power forecasting, and medical signal processing.

July, 2025

Jun Wang and *Hong Peng*
Xihua University, Chengdu, China

Acknowledgments

This book is the result of the collaboration between the authors and multiple graduate students. These graduate students are responsible for tasks such as algorithm design, programming implementation, and experimental verification. They are Qian Liu and Lifan Long (2020–2023), Xin Xiong and Yujie Zhang (2021–2024), Min Wu, Lin Guo, and Yunzhu Gao (2022–2025), and Juan He from Grade 2023. At the same time, we would like to express our gratitude to several members of the Natural Computing Research Group at the University of Seville in Spain for their long-term collaborative research on the topic of spiking neural P systems and time series analysis, including Professor Mario J. Pérez-Jiménez, Professor Agustín Riscos-Núñez, Dr. David Orellana-Martín, and Dr. Antonio Ramírez-de-Arellano.

The research work of this book was completed with support from the National Natural Science Foundation of China (Nos. 62176216 and 62076206) and the Sichuan Provincial Science Foundation Project (No. 2022ZYD0115). We would like to express our gratitude.

Acronyms

AI	Artificial Intelligence
ARIMA	Autoregressive Integrated Moving Average
BPTT	Backpropagation Through Time
ANN	Artificial Neural Networks
CNN	Convolutional Neural Networks
CNP	Coupled Neural P (systems)
CUDA	Compute Unified Device Architecture
DE	Differential Evolution
DNA	Deoxyribonucleic Acid
DTNP	Dynamic Threshold Neural P (systems)
EFSNP	Echo-like Feedback Spiking Neural P (model)
EMD	Empirical Mode Decomposition
ENSNP	Echo-like Nonlinear Spiking Neural P (model)
ESN	Echo State Networks
ESNP	Echo-like Spiking Neural P (model)
GA	Genetic Algorithm
GPU	Graphics Processing Unit
GRU	Gated Recurrent Unit
GNSNP	Gated Nonlinear Spiking Neural P (model)
GSNP	Gated Spiking Neural P (model)
IMF	Intrinsic Mode Functions
LSTM	Long Short-term Memory network
LSTM-SNP	Long Short-term Memory-like Spiking Neural P (model)
MTS	Multivariate Time Series
NSNP	Nonlinear Spiking Neural P (systems)

NSNP-AU	Nonlinear Spiking Neural P (systems) with Autapses
NSNP-TO	Nonlinear Spiking Neural P (systems) with Two Outputs
PSO	Particle Swarm Optimization
PSPACE	Polynomial SPACE Problem
RC	Reservoir Computing
RC-SNP	Reservoir Computing Based on Spiking Neural P (systems)
RMS	Reservoir Mode Space
RNN	Recurrent Neural Networks
SAT	Boolean Satisfiability Problem
SNP	Spiking Neural P (systems)
SNP-AU	Spiking Neural P (systems) with Autapses
SVM	Support Vector Machine
TSC	Time Series Classification
UTS	Univariate Time Series

1

Introduction

1.1 Background

Natural computing is a computing paradigm abstracted from nature, aimed at imitating the inherent phenomena and laws of nature or human society, exploring new algorithms for problems, and better solving them [1]. Natural computing typically encompasses adaptive computational frameworks capable of autonomous organization and continuous improvement, addressing intricate challenges beyond conventional computing paradigms. So far, natural computing has proposed numerous algorithms corresponding to various origins of inspiration, such as swarm intelligence algorithms (biological populations), artificial life (biological individuals), immune algorithms (immune systems), artificial neural networks (biological nervous systems), membrane computing (living cells), and DNA computing (DNA molecules), as well as some methods derived from physical and chemical phenomena.

Proposed by Gh. Păun in 1998, membrane computing represents an innovative computational framework within the broader field of natural computing [2]. Membrane computing aims to abstract computing models from the structure and function of living cells as well as tissues and organs that collaborate to process information. Membrane systems, alternatively called P systems, constitute a class of inherently parallel and spatially distributed computational models in membrane computing. In P systems, the interior of the cell is divided into several regions separated by the cell membranes, which are considered independent computing units. Each region contains multisets of objects (corresponding to compounds that are constantly evolving inside the cell) as carriers of information, and each computational module's handling of object multisets (analogous to cellular biochemical reactions) constitutes its autonomous information processing mechanism. Intercellular object transfer across membranes embodies the computational units' information-sharing mechanism. With distributed processing

Spiking Neural P Systems for Time Series Analysis, First Edition. Jun Wang and Hong Peng.

elements working in parallel, P systems achieve significantly greater theoretical efficiency than von Neumann-type computers [3]. Their computational power has been mathematically proven equivalent to Turing machines, with indications of transcending these classical constraints.

Drawing upon the electrophysiological properties of neural signal transmission, Ref. [4] introduced spiking neural P systems (SNP systems) in 2006 – a novel membrane computing framework mimicking neuronal architectures. The structural foundation of SNP systems consists of directed graphs wherein neuronal units serve as computational nodes interconnected by oriented synaptic pathways, facilitating concurrent distributed processing. For each neuron, there exists a state variable representing the number of spikes within it (simulating the membrane potential of biological neuron). Moreover, a set of spiking/firing rules is defined to update the state of the neuron (simulating the growth or decay of membrane potential). Due to common features with spiking neural networks [5], SNP systems are classified among third-generation neural network architectures. These research results have demonstrated that SNP systems can achieve Turing-complete computational power in many cases and have also demonstrated excellent performance and potential in solving computationally difficult problems.

Within the domain of biologically-inspired computation, membrane computing represents a significant advancement, offering fresh modeling techniques and computational methodologies for computing research. Concurrently, it furnishes innovative theoretical frameworks, methodological advancements, and practical instruments to address real-world challenges.

1.2 Membrane Computing

During his 1998 academic stay in Finland, Gh. Păun initially formulated the principles of membrane computing, which were subsequently documented in his pioneering 2000 publication [2]. Numerous researchers have focused their attention on membrane computing because of its extensive potential for real-world applications in diverse fields such as computer science, systems biology, and linguistics. The Institute for Scientific Information (ISI) recognized membrane computing in 2003 as an emerging and fast-growing discipline within computer science. The pioneering membrane computing paper emerged as both an immediate scientific breakthrough and one of ISI's most frequently referenced publications during its publication year. Springer introduced the field's first comprehensive publication on membrane computing during 2002, titled "*Membrane Computing – An Introduction*" [6]. Oxford University Press published "*The Oxford Handbook of Membrane Computing*" in 2010 [3]. In 2017, Springer published the book "*Real Life Application with Membrane Computing*" [7]. In 2024, Springer published the book "*Advanced Spiking Neural P Systems: Models and*

Applications" [8]. The membrane computing domain has yielded substantial academic output to date, including 40+ journal special issues, 50+ conference proceedings volumes, 70+ doctoral dissertations, and numerous review articles.

In membrane computing community, theoretical research, application research, and software and hardware implementation are the three main contents. Below, we will briefly review their research status.

1.2.1 Theoretical Research

With over two decades of continuous advancement, membrane computing has established a substantial theoretical foundation. P systems exhibit three primary architectural variants based on their topological configurations: cell-like (hierarchical organization), tissue-like (networked arrangement), and neural-like (directed graph architecture). The biological diversity observed in living cells has led to the creation of numerous P system architectures, exemplified by catalytic variants, membrane transport systems utilizing symport/antiport mechanisms, and promoter/inhibitor-regulated models [9, 10]. Drawing on fundamental concepts in computer science, P systems exhibiting varied operational mechanisms and parallel processing capabilities have been constructed [11]. By simulating with computing devices in classical computational theory, including Turing machines, registration machines, etc., it was confirmed that most P systems are Turing complete, meaning that the computing power of these systems is equivalent to that of Turing machines. By introducing mechanisms such as cell division, cell separation, and cell generation, P systems can generate their own computational space during the computation process. This allows these systems to effectively solve some classic NP-problems, such as PSPACE, knapsack, and SAT problems, by exchanging space for time [12, 13].

1.2.2 Application Research

Due to some characteristics of membrane computing, such as distribution, uncertainty, and scalability, the computational framework of P systems demonstrates particular efficacy in resolving specific real-world problems. In the early stages of applications, the focus was mainly on the research of membrane algorithms. The fusion between the naturally parallelized framework of P systems and established optimization approaches (genetic algorithm [GA], particle swarm optimization [PSO], differential evolution [DE]) gave rise to an innovative algorithmic paradigm. The membrane algorithm was initially developed by Ref. [14] as a computational approach for combinatorial optimization challenges, with its practical implementation demonstrated through successful application to the traveling salesman problem. The integration of membrane computing with established optimization techniques has spawned multiple algorithmic derivatives, demonstrating effectiveness across diverse problem domains ranging from theoretical models to real-world applications. Beyond its core applications, membrane computing demonstrates significant utility across interdisciplinary domains including ecological system simulations [15], visual data analysis [16], and intelligent algorithm design [17].

In recent years, membrane computing learning models have received attention, focusing on how to construct learning models using the mechanisms of membrane computing. The unsupervised learning model for membrane computing focuses on clustering algorithms. By utilizing the structures of different P systems and the mechanisms of cells (or membranes), some membrane inspired clustering algorithms have been proposed [18–22], known as membrane clustering algorithms. These algorithms discussed problems such as data clustering, fuzzy clustering, multi-objective clustering, and automatic clustering.

1.2.3 Software and Hardware Implementation

With the deepening of research, the software and hardware implementation of membrane computing has made certain progress. Several simulation software have been developed for various membrane systems, including Membrane Simulator based on C++ for catalytic P systems and active membrane P systems [23], and SNUPS based on Java for numerical P systems [24]. However, current P system simulation tools predominantly exhibit limited functionality, typically restricted to emulating specific membrane system configurations. A research team specializing in natural computation at Spain's Seville University developed P-Lingua, an integrative modeling framework [25]. The framework enables comprehensive P system modeling and execution through Java-based pLinguaCore kernel invocations. For hardware-level realization, the team employed CUDA parallel computing architecture to enable GPU-accelerated simulation of membrane computing systems [26] and proved that GPU parallel computing can greatly improve computational efficiency by solving difficult problems on the P system's GPU simulator.

1.3 Spiking Neural P Systems

SNP systems are computing systems derived from the way neurons in biological neural networks process information in the form of spikes and interact with each other using pulses, and were proposed by Ref. [4]. Neural computing is a hot research field at the intersection of neuroscience and computational science, and as a new type of third-generation neural computing models, SNP systems have become a research hotspot since their proposal, achieving numerous research results. Simultaneously, the neural computing research community has disseminated significant findings through high-impact journal publications, such as *Neural Networks and Neurocomputing* under Elsevier, *International Journal of Neural Systems* under World Scientific, and *IEEE Transactions on Neural Networks and Learning Systems*.

SNP systems represent a category of bio-inspired computing architectures that exhibit both parallel processing and distributed characteristics within membrane computing paradigms [8, 27]. SNP systems fundamentally comprise three key components: a directed graph architecture, neuronal units containing discrete spike counts, and operational rules governing both spike consumption and generation processes. The operational rules fundamentally determine the spiking dynamics that define SNP systems' computational behavior.

In the past decade, the investigation of SNP systems has yielded substantial findings across fundamental theory, practical applications, and software and hardware aspects. Below, we will introduce each of these three aspects separately.

1.3.1 Theoretical Research Work

The theoretical work of SNP systems mainly involves establishing various computational models and analyzing their computational capabilities, including two key aspects: Turing-equivalent computational capability assessment and the feasibility of deriving efficient algorithms for NP-hard problems.

Since Ref. [4] proposed SNP systems, numerous adaptations have been derived from diverse biological discoveries in the past decade, particularly those observed in neuroscience. Here are some discoveries listed: multiple channels [19], anti-spikes [28], astrocytes [29], polarization [30], rules on synapses [31], structural plasticity [32], communications on requests [33], inhibitory rules [34], autapses [35], dendritic mechanism [34], delay on synapses [36], nonlinear mechanism [37], coupling modulation [38], dynamic threshold [39], plasmids [40], etc. Recent studies on SNP systems reveal an increasing number of biologically inspired mechanisms.

The first theoretical question to address after building the model is its computational completeness, specifically examining if it achieves the same level of computation as a Turing machine. When examining computational completeness, computational models to analyze encompass digital generators/acceptors, functional computing devices, and formal language generators. Many results demonstrate that these models exhibit Turing-complete capabilities, such as digital generators/acceptors and functional computing devices [3]. In addition, it has been proven that some variants can generate regular languages and recursively enumeration languages [41].

SNP models generally operate under three distinct execution modes: sequential, asynchronous, and synchronous. For synchronous case, a global clock synchronizes each neuron. Therefore, neurons operate concurrently, however, every neuron uses the rules in sequence. In previous variants, most work in synchronous case. For asynchronous case, all neurons work asynchronously without considering the global clock [42]. The past few years have seen theoretical verification of computation universality in some asynchronous models. For sequential case, processing scheme requires neurons to activate one after another, while maintaining rule execution order within individual processing units [43]. Researchers have introduced various sequential models that demonstrate computational universality, serving as both digital generators/acceptors and functional computing devices.

Computational theorists continue developing scaled-down universal systems, exemplified by space-efficient Turing machines and register machines with minimal instruction sets. The investigation of minimal universal models has become an active research direction in SNP systems, with Gh. Păun pioneering the foundational work on their computational completeness at small scales [29].

Another key theoretical focus involves analyzing the computational efficiency of SNP system architectures. This study aims to evaluate computational efficiency through the application of SNP architectures and their modified forms in addressing some NP-hard problems. The purpose is to test their efficiency by using SNP systems and variants to solve some NP-hard problems, including PSPACE and SAT problems [44]. SNP systems show great potential in solving these NP-hard problems. Currently, multiple approaches exist for addressing this challenge: (i) Using exponential scale precomputing resources (i.e., the system has exponential neurons) to solve the NP-hard problem through spatial exchange time; (ii) The polynomial scale SNP systems can solve NP-hard problems in a nondeterministic manner in polynomial time; (iii) Enhancing neuronal density in computational architectures facilitates a space-for-time exchange strategy, which proves effective against NP-complete challenges.

1.3.2 Model Application Work

With their distributed framework, massive parallel processing capacity, and tolerance for ambiguous dynamics, SNP systems demonstrate unique advantages in handling nonlinear real-life challenges that require flexible computational approaches.

The computational properties of SNP systems have enabled the development of diverse hardware components [45], including elementary logic elements, digital circuit implementations, natural number arithmetic units, and probabilistic computing modules.

Researchers have employed the nondeterministic nature of SNP systems to create enhanced algorithmic solutions [46], where refined SNP architectures address complex combinatorial optimization challenges.

The incorporation of fuzzy set theory into SNP architectures has led to the creation of new computational models with enhanced uncertainty handling capabilities. Subsequently, these enhanced SNP models with fuzzy capabilities found practical implementation in diagnosing faults within electrical transformers and broader power grid systems [47, 48].

Nonlinear spiking neural P (NSNP) systems, coupled neural P (CNP) systems, and dynamic threshold neural P (DTNP) systems were applied in multiple image processing tasks, including image fusion [49], image segmentation [50], and edge detection [51, 52].

The nonlinear spiking mechanism of NSNP systems has been used to develop several recurrent-like models and convolutional models, including long- and short-term memory-like spiking neural P systems (or LSTM-SNP model) [53], nonlinear spiking neural P systems with autapses (or NSNP-AU model) [54], gated spiking neural P systems (or GSNP model) [55], and ConvSNP model [56]. These models were applied in time series [8, 57], image processing [52, 58], sentiment analysis [59, 60], and recommendation systems [61, 62].

1.3.3 Hardware and Software Simulation

Significant advancements have been achieved in creating computational tools for simulating SNP systems. As an illustration, Ref. [63] developed computational models within the P-Lingua environment to simulate both conventional and nonsynchronous SNP systems. Ref. [64] created Pnet Lab, a computational platform designed for the analysis and emulation of both standard SNP systems and their anti-spike variants. Furthermore, Ref. [65] engineered an online interactive platform capable of graphically simulating SNP systems.

From a computational hardware perspective, studies have examined how matrix formulations can represent spiking neural membrane computing models [66]. Building upon the matrix model, developers engineered a high-performance simulation platform harnessing GPU parallel processing through CUDA programming. The massively parallel design of GPUs aligns well with the concurrent processing nature of SNP systems, enabling significant enhancements in simulation speed through hardware acceleration.

1.4 Time Series Analysis

Today, we have entered the digital age and data is growing explosively. As a collection of data arranged in chronological order, time series data is widely present in various fields, such as stock price trends in financial markets, temperature changes in meteorological fields, and equipment operating parameters in industrial production. Conducting in-depth analysis of time series data and uncovering the underlying patterns and trends is of crucial importance for decision-making, risk prediction, and other related areas [67].

As a multidisciplinary field, machine learning aims to enable computers to learn features and patterns through data, thereby achieving tasks such as predicting and classifying unknown data. As a subset of machine learning, deep learning employs multi-layered neural architectures to autonomously extract intricate patterns from extensive datasets. Advancements in computational capabilities have propelled machine and deep learning to the forefront of temporal data studies, where their efficacy continues to attract significant academic attention.

Classical approaches, autoregressive moving average (ARMA) and exponential smoothing, perform well in handling simple linear time series data, but often struggle to achieve ideal results for time series data with complex nonlinear relationships and long-term dependencies. Machine learning and deep learning models, with their powerful nonlinear modeling and automatic feature extraction capabilities, can effectively capture complex dynamic features and patterns in time series data, bringing a new solution for time series analysis.

In the financial field, forecasting accurate time series can help investors grasp market trends, formulate reasonable investment strategies, reduce risks, and increase returns. For example, by analyzing and predicting the time series of stock prices, investors can decide when to buy or sell stocks. In industrial production, time series classification can be used for equipment fault diagnosis and quality control. By analyzing the time series data of equipment operation status, the normal operation status and fault status are classified, and potential fault hazards are discovered in a timely manner to ensure the continuity and stability of production. Anomaly detection can identify abnormal data points in the production process, take timely measures to adjust, and avoid production accidents. In the field of transportation, by analyzing and predicting the time series of traffic flow, it is possible to plan traffic resources reasonably, optimize traffic signal timing, and alleviate traffic congestion.

1.4.1 Prediction Problems

Forecasting future data points by analyzing historical temporal sequences is a widely utilized technique across multiple disciplines, such as finance, economics, energy, meteorology, and more [68]. Accurate time series forecasting can help companies develop reasonable production plans, investors make wise investment decisions, and governments carry out effective policy planning. With the continuous development of techniques, many methods for time series forecasting have emerged, which can be roughly divided into machine learning-based and deep learning-based approaches [69].

Classical prediction approaches mainly include autoregressive moving average (ARIMA) and their extensions, such as seasonal autoregressive moving average (SARIMA) model, and exponential smoothing methods. However, the ARIMA model also has some limitations. It requires high stationarity of data, and for non-stationary time series data, complex differencing is required. Improper selection of different orders may affect the performance of the model. The ARIMA model assumes that time series data have a linear relationship, and for time series with complex nonlinear relationships, its prediction performance is often not ideal.

As deep neural architectures continue to evolve, their utilization for forecasting sequential data patterns has expanded significantly in practical implementations. Deep learning models have powerful nonlinear modeling and feature extraction capabilities, which can effectively process complex time series data, capturing long-term dependency, and sequential pattern.

Common deep learning models used for time series prediction include (i) gated recurrent unit (GRU) and long short-term memory network (LSTM); (ii) convolutional neural network (CNN); (iii) transformer model. Deep learning models such as LSTM, GRU, transformer, and their variants typically exhibit better performance and can more accurately obtain complex patterns and trends in data [69, 70]. CNN related models have advantages in capturing local features of time series and are suitable for processing data with obvious local patterns and trends.

1.4.2 Classification Problems

The classification task is to assign time series data to predefined categories and has wide applications in fields such as industrial manufacturing, medical diagnosis, and financial risk assessment [71]. For example, in industrial manufacturing, by classifying the time series data of equipment operation, it is possible to determine whether the equipment is in a normal operating state. In medical diagnosis, based on the patient's physiological indicators and time series data, it can assist doctors in diagnosing and classifying diseases.

Machine learning has also been extensively applied in time series classification. Traditional machine learning models include support vector machine (SVM) [72] and random forest (RF) [73].

Deep learning methods can automatically capture complex features in time series, avoiding the tedious process of manual feature extraction, and have advantages in processing large-scale data [71]. These deep learning methods include (i) RNN, LSTM, and GRU models; (ii) CNN-based models; (iii) transformer-based models.

1.4.3 Anomaly Detection Problems

The anomaly detection for time series is to identify data samples or subsequences in time series data that do not conform to normal patterns [74]. It has important application value in financial risk warning, industrial equipment fault diagnosis, network security monitoring, and other fields. For example, in the financial sector, timely detection of abnormal fluctuations in stock prices can help investors avoid significant losses. In industrial production, detecting abnormal changes in equipment operating parameters can prevent equipment failures in advance, ensuring the continuity and safety of production. With the development of machine learning and deep learning technologies, time series anomaly detection methods are constantly innovating and improving.

Statistical methods are commonly used traditional methods in time series anomaly detection, mainly based on the statistical features of data to identify outliers. Moving average is a simple and commonly used method that identifies outliers by calculating the average around a data point.

Machine learning models were widely applied in time series anomaly detection, capable of handling more complex data patterns and relationships, including isolation forests (IF) [75] and one class SVM [76].

Deep learning methods have demonstrated powerful capabilities in time series anomaly detection [77], capable of automatically learning complex features and patterns of data. These deep learning methods include AutoEncoder, LSTM, etc. They demonstrate powerful capabilities in fields such as power system fault detection and aircraft engine health monitoring by learning patterns and patterns from normal data, utilizing reconstruction errors or prediction biases to detect anomalies.

1.5 Organization of Chapters

Organized in three parts, this book comprises eight chapters in total. As a foundation, the first part establishes the core concepts, including Chapters 1 and 2. Serving as the introductory section, Chapter 1 initially explores the scholarly context and membrane computation before delving into the developmental milestones of spiking neural P systems and time series analysis. Chapter 2 describes spiking neural P systems, and then discusses three theoretical variants, including spiking neural P systems with extended rules, spiking neural P systems with autapses, and nonlinear spiking neural P systems.

The second part introduces three classes of application models of time series analysis, including three chapters. Chapter 3 discusses the recurrent-like models for time series prediction. Chapter 4 discusses the echo-like models for time series prediction. Chapter 5 discusses the reservoir computing models for time series classification.

Comprising three chapters, the final part focuses on practical applications. Chapter 6 introduces the applications in financial time series analysis. Chapter 7 introduces the applications in power load forecasting and photovoltaic power forecasting tasks. Focusing on medical signal processing problem, Chapter 8 presents corresponding method.

References

1 D.E. Shasha and C. Lazere. *Natural Computing*. W. W. Norton & Company, 2010.

2 Gh. Păun. Computing with membranes. *Journal of Computer and System Sciences*, 61(1): 108–143, 2000.

3 Gh. Păun, G. Rozenberg, and A. Salomaa. *The Oxford Handbook of Membrane Computing*. Oxford University Press, New York, 2010.

4 M. Ionescu, Gh. Păun, and T. Yokomori. Spiking neural P systems. *Fundamenta Informaticae*, 71: 279–308, 2006.

5 S. Ghosh-Dastidar and H. Adeli. Spiking neural networks. *International Journal of Neural Systems*, 19(4): 295–308, 2009.

6 Gh. Păun. *Membrane Computing: an Introduction*. Springer-Verlag, Berlin, Heidelberg, 2002.

7 G. Zhang, M.J. Pérez-Jiménez, and M. Gheorghe. *Real-Life Application with Membrane Computing*. Springer, 2017.

8 H. Peng and J. Wang. *Advanced Spiking Neural P Systems: Models and Applications*. Springer, Singapore, 2024.

9 A. Păun and Gh. Păun. The power of communication: P systems with symport/antiport. *New Generation Computing*, 20(3): 295–305, 2002.

10 F. Bernardini and M. Gheorghe. Languages generated by P systems with active membranes. *New Generation Computing*, 22(4): 311–329, 2004.

11 R. Freund. Asynchronous P systems and P systems working in the sequential mode. *Lecture Notes in Computer Science*, 3365: 36–62, 2004.

12 P. Sośık, A. Păun, and A. Rodríguez-Patón. P systems with proteins on membranes characterize PSPACE. *Theoretical Computer Science*, 488: 78–95, 2013.

13 B. Song, M.J. Pérez-Jiménez, and L. Pan. Efficient solutions to hard computational problems by P systems with symport/antiport rules and membrane division. *BioSystems*, 130: 51–58, 2015.

14 T.Y. Nishida. An application of P systems: a new algorithm for NP-complete optimization problems. *Proceedings of the 8th World Multi-Conference on Systems, Cybernetics and Informatics*, pages 109–112, 2004.

15 M. García-Quismondo, M. Levin, and D. Lobo. Modeling regenerative processes with membrane computing. *Information Sciences*, 381: 229–249, 2017.

16 D. Díaz-Pernil, M.A. Gutiérrez-Naranjo, and H. Peng. Membrane computing and image processing: a short survey. *Journal of Membrane Computing*, 1: 58–73, 2019.

17 C. Buiu and A.G. Florea. Membrane computing models and robot controller design, current results and challenges. *Journal of Membrane Computing*, 1(4): 262–269, 2019.

18 J. Yang, H. Peng, X. Luo, and J. Wang. Stochastic numerical P systems with application in data clustering problems. *IEEE Access*, 8(1): 31507–31518, 2020.

19 H. Peng, P. Shi, J. Wang, A. Riscos-Núñez, and M.J. Pérez-Jiménez. Multiobjective fuzzy clustering approach based on tissue-like membrane systems. *Knowledge-Based Systems*, 125: 74–82, 2017.

20 H. Peng, J. Wang, P. Shi, M.J. Pérez-Jiménez, and A. Riscos-Núñez. An extended membrane system with active membrane to solve automatic fuzzy clustering problems. *International Journal of Neural Systems*, 26(3): 1650004, 2016.

21 H. Peng, J. Wang, P. Shi, A. Riscos-Núñez, and M.J. Pérez-Jiménez. An automatic clustering algorithm inspired by membrane computing. *Pattern Recognition Letters*, 68: 34–40, 2015.

22 H. Peng, J. Wang, M.J. Pérez-Jiménez, and A. Riscos-Núñez. An unsupervised learning algorithm for membrane computing. *Information Sciences*, 304: 80–91, 2015.

23 G. Ciobanu and D. Paraschiv. P system software simulator. *Fundamenta Informaticae*, 49(1–3): 61–66, 2002.

24 O. Arsene, C. Buiu, and N. Popescu. SNUPS–a simulator for numerical membrane computing. *International Journal of Innovative Computing, Information and Control*, 7(6): 3509–3522, 2011.

25 D. Díaz-Pernil, I. Pérez-Hurtado, M.J. Pérez-Jiménez, and A. Riscos-Núñez. A P-lingua programming environment for membrane computing. *Lecture Notes in Computer Science*, 5391: 187–203, 2008.

26 J.M. Cecilia, García J.M., G.D. Guerrero, M.A. Martínez-del-Amor, P.H. Ignacio, and M.J. Pérez-Jiménez. Simulation of P systems with active membranes on CUDA. *Briefings in Bioinformatics*, 11(3): 313–322, 2009.

27 H. Chen, M. Ionescu, T.O. Ishdorj, A. Păun, Gh. Păun, and MJ. Pérez-Jiménez. Spiking neural P systems with extended rules: universality and languages. *Natural Computing*, 7: 147–166, 2008.

28 L. Pan and Gh. Păun. Spiking neural P systems with anti-spikes. *International Journal of Computers, Communications & Control*, 4(3): 273–282, 2009.

29 Gh. Păun. Spiking neural P systems with astrocyte-like control. *Journal of Universal Computer Science*, 13(11): 1707–1721, 2007.

30 T. Wu, A. Păun, Z. Zhang, and L. Pan. Spiking neural P systems with polarizations. *IEEE Transactions on Neural Networks and Learning Systems*, 29(8): 3349–3360, 2017.

31 T. Song, L. Pan, and Gh. Păun. Spiking neural P systems with rules on synapses. *Theoretical Computer Science*, 529: 82–95, 2014.

32 R.T.A. de la Cruz, F.G.C. Cabarle, I.C.H. Macababayao, H.N. Adorna, and X. Zeng. Homogeneous spiking neural P systems with structural plasticity. *Journal of Membrane Computing*, 3(1): 10–21, 2021.

33 L. Pan, Gh. Păun, G. Zhang, and F. Neri. Spiking neural P systems with communication on request. *International Journal of Neural Systems*, 28(8): 1750042, 2017.

34 H. Peng, B. Li, J. Wang, X. Song, T. Wang, L. Valencia-Cabrera, I. Pérez-Hurtado, A. Riscos-Núñez, and M.J. Pérez-Jiménez. Spiking neural P systems with inhibitory rules. *Knowledge-Based Systems*, 30(2): 2050008, 2020.

35 X. Song, L. Valencia-Cabrera, H. Peng, and J. Wang. Spiking neural P systems with autapses. *Information Sciences*, 570: 383–402, 2021.

36 X. Song, L. Valencia-Cabrera, H. Peng, J. Wang, and M.J. Pérez-Jiménez. Spiking neural P systems with delay on synapses. *International Journal of Neural Systems*, 31(1): 2050042, 2021.

37 H. Peng, Z. Lv, B. Li, X. Luo, J. Wang, X. Song, T. Wang, M.J. Pérez-Jiménez, and A. Riscos-Núñez. Nonlinear spiking neural P systems. *International Journal of Neural Systems*, 30(10): 2050008, 2020.

38 H. Peng and J. Wang. Coupled neural P systems. *IEEE Transactions on Neural Networks and Learning Systems*, 30(6): 1672–1682, 2019.

39 H. Peng, J. Wang, M.J. Pérenz-Jiménez, and A. Riscos-Núñez. Dynamic threshold neural P systems. *Knowledge-Based Systems*, 163: 875–884, 2019.

40 F.G.C. Cabarle, X. Zeng, N. Murph, T. Son, A. Rodríguez-Patón, and X. Liu. Neural-like P systems with plasmids. *Information and Computation*, 281(4): 104766, 2021.

41 Y. Huang, W. Yi, H. Peng, J. Wang, X. Luo, and Q. Yang. Computational power of dynamic threshold neural P systems for generating string languages. *Theoretical Computer Science*, 851: 77–91, 2021.

42 X. Song, J. Wang, H. Peng, G. Ning, Z. Sun, T. Wang, and F. Yang. Small universal asynchronous spiking neural P systems with multiple channels. *Neurocomputing*, 378: 1–8, 2020.

43 T. Bao, N. Zhou, H. Peng, Q. Yang, and J. Wang. Computational completeness of sequential spiking neural P systems with inhibitory rules. *Information and Computation*, 281(1): 104786, 2021.

44 L. Pan, Gh. Păun, and MJ. Pérez-Jiménez. Spiking neural P systems with neuron division and budding. *Science China Information Sciences*, 54(8): 1596–1607, 2011.

45 T. Song, P. Zheng, M.D. Wong, and X. Wang. Design of logic gates using spiking neural P systems with homogeneous neurons and astrocytes-like control. *Information Sciences*, 372: 380–391, 2016.

46 M. Zhu, Q. Yang, J. Dong, G. Zhang, X. Gou, H. Rong, P. Paul, and F. Neri. An adaptive optimization spiking neural P systems for binary problems. *International Journal of Neural Systems*, 31(1): 1440006, 2021.

47 H. Peng, J. Wang, J. Ming, P. Shi, M.J. Pérez-Jiménez, W. Yu, and C. Tao. Fault diagnosis of power systems using intuitionistic fuzzy spiking neural P systems. *IEEE Transaction on Smart Grid*, 9(5): 4777–4784, 2018.

48 J. Wang, H. Peng, W. Yu, M. Ming, M.J. Pérez-Jiménez, C. Tao, and X. Huang. Interval valued fuzzy spiking neural P systems for fault diagnosis of power transmission networks. *Engineering Applications of Artificial Intelligence*, 82: 102–109, 2019.

49 B. Li, H. Peng, J. Wang, and X. Huang. Multi-focus image fusion based on dynamic threshold neural P systems and surfacelet transform. *Knowledge-Based Systems*, 196: 105794, 2020.

50 Y. Cai, S. Mi, J. Yan, H. Peng, X. Luo, Q. Yang, and J. Wang. An unsupervised segmentation method based on dynamic threshold neural P systems for color images. *Information Sciences*, 587: 473–484, 2022.

51 R. Xian, R. Lugu, H. Peng, Q. Yang, X. Luo, and J. Wang. Edge detection method based on nonlinear spiking neural systems. *International Journal of Neural Systems*, 33(1): 2250060, 2023.

52 R. Xian, X. Xiong, H. Peng, J. Wang, A. Ramírez-de-Arellano, and Q. Yang. Feature fusion method based on spiking neural convolutional network for edge detection. *Pattern Recognition*, 147: 110112, 2024.

53 Q. Liu, L. Long, Q. Yang, H. Peng, J. Wang, and X. Luo. LSTM-SNP: A long short-term memory model inspired from spiking neural P systems. *Knowledge-Based Systems*, 235: 107656, 2022.

54 Q. Liu, H. Peng, L. Long, J. Wang, Q. Yang, M.J. Pérez-Jiménez, and D. Orellana-Martín. Nonlinear spiking neural systems with autapses for predicting chaotic time series. *IEEE Transactions on Cybernetics*, 54(3): 1841–1853, 2024.

55 Q. Liu, L. Long, H. Peng, J. Wang, Q. Yang, X. Song, A. Riscos-Núñez, and M.J. Pérez-Jiménez. Gated spiking neural P systems for time series forecasting. *IEEE Transactions on Neural Networks and Learning Systems*, 34(9): 6227–6236, 2023.

56 S. Zhao, L. Zhang, H. Peng, Z. Liu, and J. Wang. ConvSNP: a deep learning model embedded with SNP-like neurons. *Journal of Membrane Computing*, 4: 87–95, 2022.

57 Y. Zhang, Q. Yang, Z. Liu, H. Peng, and J. Wang. A prediction model based on gated nonlinear spiking neural system. *International Journal of Neural Systems*, 33(6): 2350029, 2023.

58 L. Ye, C. Zhou, H. Peng, J. Wang, Z. Liu, and Q. Yang. Multi-level feature interaction image super-resolution network based on convolutional nonlinear spiking neural model. *Neural Networks*, 177: 106366, 2024.

59 Y. Huang, Q. Liu, H. Peng, J. Wang, Q. Yang, and D. Orellana-Martín. Sentiment classification using bidirectional LSTM-SNP model and attention mechanism. *Expert Systems with Applications*, 221: 119730, 2023.

60 Y. Huang, H. Peng, Q. Liu, Q. Yang, J. Wang, D. Orellana-Martín, and M.J. Pérez-Jiménez. Attention-enabled gated spiking neural P model for aspect-level sentiment classification. *Neural Networks*, 157: 437–443, 2023.

61 X. Bai, Y. Huang, H. Peng, J. Wang, Q. Yang, D. Orellana-Martín, A. Ramírez-de-Arellano, and M.J. Pérez-Jiménez. Sequence recommendation using multi-level self-attention network with gated spiking neural P systems. *Information Sciences*, 656: 119916, 2024.

62 X. Bai, L. Zhang, M. Jiang, H. Peng, J. Wang, Q. Yang, and A. Ramírez-de-Arellano. Gated graph spiking neural P network for session-based recommendation. *Knowledge-Based Systems*, 300: 112162, 2024.

63 L.F. Macías-Ramos, M.J. Pérez-Jiménez, T. Song, and L. Pan. Extending simulation of asynchronous spiking neural P systems in P-Lingua. *Fundamenta Informaticae*, 36(3): 253–267, 2015.

64 V.P. Metta, K. Krithivasan, and D. Garg. Extending simulation of asynchronous spiking neural P systems in P-Lingua. *Proceedings of the 12th International Conference on Membrane Computing*, pages 381–394, 2011.

65 A.G.S. Dupaya, A.C.A.P. Galano, F.G.C. Cabarle, R.T.D. La Cruz, K.J. Ballesteros, and P.P. Lazo. A web-based visual simulator for spiking neural P systems. *Journal of Membrane Computing*, 4: 21–40, 2022.

66 Z.B. Jimenez, F.G.C. Cabarle, R.T.A. de la Cru, K.C. Buño, H.N. Adorna, N.H.S. Hernandez, and X. Zeng. Matrix representation and simulation algorithm of spiking neural P systems with structural plasticity. *Journal of Membrane Computing*, 1: 145–160, 2019.

67 G.E. Box, G.M. Jenkins, G.C. Reinsel, and G.M. Ljung. *Time Series Analysis: Forecasting and Control*. John Wiley & Sons, 2015.

68 N. Wu and P. Smyth. Autoformer: Decomposition-based auto-regression for long-term series forecasting. *Advances in Neural Information Processing Systems*, 33: 16704–16715, 2020.

69 H. Zhou, S. Zhang, Y. Zhang, and Y. Wang. Pyraformer: pyramid-attention-based deep learning model for long-range time series forecasting. *Proceedings of the AAAI Conference on Artificial Intelligence*, 35(12): 10373–10381, 2021.

70 Y. Wang, Y. Wang, Y. Zhang, and Y. Wang. TiDE: A time series dense encoder for fast and accurate long-term time series forecasting. *Proceedings of the AAAI Conference on Artificial Intelligence*, 37(5): 5474–5483, 2023.

71 L. Zhao and H. Lu. Convolutional neural networks for time series classification: a survey. *Knowledge-Based Systems*, 169: 1–16, 2019.

72 C. Cortes and V. Vapnik. Support-vector networks. *Machine Learning*, 20(3): 273–297, 1995.

73 L. Breiman. Random forests. *Machine Learning*, 45(1): 5–32, 2001.
74 V. Chandola, A. Banerjee, and V. Kumar. Anomaly detection: a survey. *ACM computing surveys (CSUR)*, 41(3): 15, 2009.
75 F.T. Liu, K.M. Ting, and Z.H. Zhou. Isolation forest. *2008 Eighth IEEE International Conference on Data Mining*, pages 413–422, 2008.
76 J. Park and J. Kim. Anomaly detection in multivariate time series using one-class support vector machines with relative density. *Entropy*, 21(3): 248, 2019.
77 Y. Sun and X. He. A survey on deep learning for time series anomaly detection. *Neurocomputing*, 415: 100–110, 2020.

2

Spiking Neural P Systems

2.1 Introduction

As a bio-inspired computing paradigm, spiking neural P systems (SNP systems) [1] emulate the spike-based communication observed in biological neurons [2]. From a topological perspective, SNP systems can be represented through directed graphs, with neurons serving as vertices and synaptic links depicted as directed edges. These systems comprise two fundamental components: (i) the quantity of spikes present within neurons, and (ii) the operational rules governing spike consumption and generation. The collective spike counts across all neurons define the system's current configuration. The behavioral dynamics of SNP systems emerge from the application of spiking and forgetting rules, making them suitable for modeling dynamic processes.

The past decade has witnessed diverse adaptations of SNP systems emerging through interdisciplinary convergence, incorporating principles from life sciences, brain research, and digital computation. For example, some variants are listed as follows: SNP systems with astrocytes [3], SNP systems with anti-spikes [4], SNP systems with rules on synapses [5], SNP systems with polarization [6], SNP systems with plasmids [7], SNP systems with communication on request [8], SNP systems with inhibition rules [9], SNP systems with multiple channels [10], SNP systems with enzymes [11], dynamic threshold neural P systems [12], coupled neural P systems [13], SNP systems with delay on synapses [14], dendrite P systems [15]. Within computational theory frameworks, these modified systems demonstrate Turing-complete capabilities for both numerical processing (generation/recognition) and functional computation. Additionally, certain variants exhibit equivalent universality in formal language generation. More variants can be found in papers recently published in Journal of Membrane Computing and other Neural Computing journals.

Spiking Neural P Systems for Time Series Analysis, First Edition. Jun Wang and Hong Peng.

This chapter will introduce (standard) SNP systems [1] and three variants. The three variants are SNP systems with extended rules [16], SNP systems with autapses [17], and nonlinear SNP systems [18]. Every variant is initially defined, followed by an explanation of its operational principles, culminating in a practical example.

2.2 Preliminaries

SNP systems and three variants discussed in this chapter are theoretical computational models. In order to facilitate understanding of their definitions, it is necessary to briefly introduce some notations and notions in basic formal language and automata theory [19, 20].

For an alphabet O, O^* denotes the set of all finite strings of symbols from O, and O^+ denotes the set of all nonempty strings over O; λ denotes the empty string, and $O^* = O^+ \cup \{\lambda\}$. When $O = \{\alpha\}$ is a singleton, then we write simply α^* and α^+ instead of $\{\alpha\}^*$ and $\{\alpha\}^+$.

Regular expressions based on alphabet O may be formed by beginning with λ and the elements of O, then applying union, concatenation, and Kleene+ operations, with parentheses employed as needed to clarify operation precedence. The definition comprises three clauses: (i) Both the empty string λ and every element α from set O qualify as regular expressions; (ii) For regular expressions F_1 and F_2 over O, their union $(F_1) \cup (F_2)$, concatenation $(F_1)(F_2)$, and positive closure $(F_1)^+$ also constitute valid regular expressions over O; (iii) No other constructs beyond these rules can be considered as regular expressions over O. For every regular expression F, we define a corresponding formal language $L(F)$ through these rules: (i) The empty string λ maps to itself $L(\lambda) = \lambda$, while any symbol α in alphabet O represents a singleton set ($L(\alpha) = \{\alpha\}$); (ii) For any two regular expressions F_1 and F_2 over O, their union operation yields $L(F_1) \cup L(F_2)$, concatenation produces $L(F_1)L(F_2)$, and the positive closure generates $L((F_1)^+) = L(F_1)^+$.

A formal language L (where L is a subset of O^*) qualifies as regular precisely when a regular expression F exists for alphabet O that generates L, i.e., $L(F) = L$.

For ease of understanding, several examples are illustrated as follows. Suppose that $O = \{\alpha, \beta\}$, several expressions over O and the corresponding languages are provided:

1. if $F = \alpha^*\beta$, then $L(F) = \{\beta, \alpha\beta, \alpha^2\beta, \alpha^3\beta, \dots\}$;
2. if $F = \alpha^+\beta^2$, then $L(F) = \{\alpha\beta^2, \alpha^2\beta^2, \alpha^3\beta^2, \dots\}$;
3. if $F = \alpha^2\beta$, then $L(F) = \{\alpha^2\beta\}$;
4. if $F = \alpha^*$, then $L(F) = \{\lambda, \alpha, \alpha^2, \alpha^3, \dots\}$;
5. if $F = \alpha^+$, then $L(F) = \{\alpha, \alpha^2, \alpha^3, \dots\}$;
6. if $F = \alpha^2$, then $L(F) = \{\alpha^2\}$;

7. if $F = \alpha^3$, then $L(F) = \{\alpha^3\}$;
8. if $F = \alpha^n$, then $L(F) = \{\alpha^n\}$;

2.3 Spiking Neural P Systems

This section presents SNP systems, a neural-like theoretical computational model, originally proposed in Ref. [1], including their definition, working mechanism, and an illustrative example. For simplicity, the delay factor d in SNP systems defined by Ref. [1], is ignored.

2.3.1 Definition

Definition 2.1 A SNP system with n neurons is given by

$$\Pi = (O, \sigma_1, \ldots, \sigma_n, syn, in, out) \tag{2.1}$$

wherein

1. $O = \{\alpha\}$ is the singleton alphabet, where α is called the spike;
2. $\sigma_1, \sigma_2, \ldots, \sigma_n$ are n neurons, of the form $\sigma_i = (v_i, R_i),\ i = 1, 2, \ldots, n$, wherein
   a. $v_i \geq 0$ is the initial number of spikes in σ_i;
   b. R_i is the set of two types of rules:
      i. Spiking/firing rule: $F/\alpha^c \to \alpha$, in which F is a regular expression over α, and $c \geq 1$;
      ii. Forgetting rule: $\alpha^s \to \lambda$ for $s \geq 1$, with the restriction that for each rule $F/\alpha^c \to \alpha$ of type (i) from R_i, we have $\alpha^s \notin L(F)$;
3. $syn \subseteq \{1, \ldots, n\} \times \{1, \ldots, n\}$ with $i \neq j$ for all $(i, j) \in syn$, $1 \leq i, j \leq n$ (synapses);
4. $out, in \in \{1, \ldots, n\}$ assign the output and input neurons, respectively.

From a topological perspective, SNP systems can be represented using directed graphs, with neurons serving as vertices and synaptic connections depicted as directed edges. Beyond this structural framework, every neuron in a SNP system possesses two key components: state and rules. The state is quantified by the accumulated spike count within the neuron. Let $v_i(\tau)$ denote the state of σ_i at timestep τ. Thus, $v_i(0) = v_i$. It is important to note that in SNP systems, the states of neurons are the integer. From the perspective of neural coding, the SNP system adopts a rate coding form.

In the SNP system, the state of each neuron is evolved by rules. Therefore, rules characterize the behavior of the entire system. There are two classes of rules: (i) firing/spiking rule and (ii) forgetting rule. The firing/spiking rules have the form $F/\alpha^c \to \alpha$, in which F denotes the regular expression with α.

The working mechanism of these spiking rules is illustrated below. Assume that σ_i has the rule $F/\alpha^c \rightarrow \alpha$. F is the regular expression with α, therefore, $L(F)$ is used to denote the language generated by F. If σ_i has v spikes at timestep τ, and $\alpha^v \in L(F)$, $v \geq c$, then rule $F/\alpha^c \rightarrow \alpha$ can be used. When neuron σ_i fires, c spikes are consumed, which means that $(v - c)$ spikes remain in it, and then it generates one spike, which is sent to its post-synaptic neurons σ_j, i.e., $(i, j) \in syn$. For the rule $F/\alpha^c \rightarrow \alpha$, if $F = \alpha^c$, then it is expressed by $\alpha^c \rightarrow \alpha$.

The working mechanism of forgetting rules is described below. If σ_i has exactly s spikes, then $\alpha^s \rightarrow \lambda$ can be applied, which means that neuron σ_i removes all s spikes.

Based on the semantics of these rules, state equation of σ_i is given by

$$v_i(\tau) = \begin{cases} v_i(\tau - 1) - c + m, & \text{if spiking rule } F/\alpha^c \rightarrow \alpha \text{ is applied;} \\ v_i(\tau - 1) - s + m, & \text{if forgetting rule } \alpha^s \rightarrow \lambda \text{ is applied;} \\ v_i(\tau - 1) + m, & \text{otherwise.} \end{cases} \tag{2.2}$$

wherein $v_i(\tau - 1)$ and $v_i(\tau)$ are the states of σ_i at timesteps $\tau - 1$ and τ, respectively; m is the received number of spikes from its presynaptic neurons; c is the consumed number of spikes, s is the removed number of spikes. As demonstrated in Eq. (2.2), the neuronal state equation takes the form of either a linear differential/discrete equation or a piecewise linear function.

During every computational step, neuron σ_i is required to apply one of its available rules when possible. The condition where $L(F_1) \cap L(F_2) \neq \emptyset$ may occur between two spiking rules $F_1/\alpha^{c_1} \rightarrow \alpha$ and $F_2/\alpha^{c_2} \rightarrow \alpha$, potentially enabling multiple rules to fire simultaneously in a single neuron. In such scenarios, the system selects one applicable rule through nondeterministic choice. When a spiking rule can be applied in a neuron, forgetting rules become inapplicable, and the converse also holds true. Consequently, within individual neurons, rules are applied sequentially with only one rule active per timestep, whereas neurons operate simultaneously across the system. Notably, the nondeterministic choice between applicable rules is crucial for achieving computational universality in SNP systems and variants.

The system's initial configuration is defined by the quantity of spikes present in neurons at initialization, represented as $C_0 = (v_1, v_2, \ldots, v_n)$. During the computation, the configuration at timestep τ is expressed by $C_\tau = (v_1(\tau), v_2(\tau), \ldots, v_n(\tau))$, where $v_i(\tau)$ is the number of spikes in σ_i at timestep τ, i.e., the state of σ_i at timestep τ, $i = 1, 2, \ldots, n$. Moreover, a transition is defined from one configuration to another. A computational process is defined as any series of transitions originating from the startup arrangement. The process terminates upon encountering a configuration where no rule can be used.

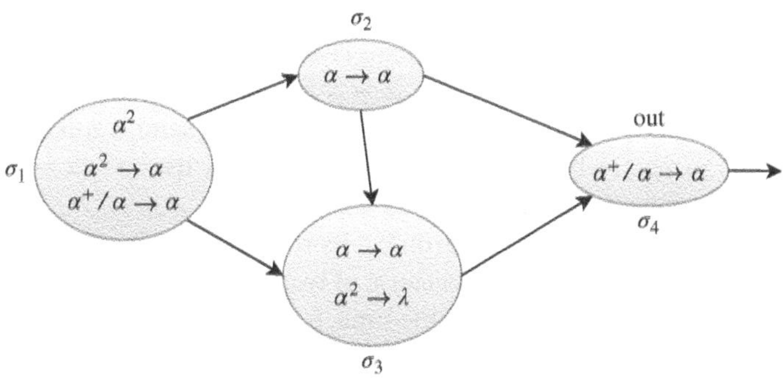

Figure 2.1 A SNP system (provided by authors).

Previous studies [1, 21] have demonstrated that SNP systems possess Turing universality, capable of operating as numerical generators/acceptors, functional computing units, and linguistic processors. However, this book does not involve the computational theory of SNP systems, but focuses on their application models.

2.3.2 Illustrative Example

The working principle of SNP systems can be observed through the elementary case presented in Figure 2.1. The SNP system contains four neurons, $\sigma_1, \sigma_2, \sigma_3$, and σ_4. And neurons σ_4 is an output neuron.

Initially, only neuron σ_1 contains two spikes, and neurons σ_2, σ_3, and σ_4 have no spikes. Thus, $C_0 = (2, 0, 0, 0)$.

At time $\tau = 1$, due to two spikes in neuron σ_1, its two rules, $\alpha^2 \rightarrow \alpha$ and $\alpha^+/\alpha \rightarrow \alpha$, can be applied. Therefore, one of them is chosen to apply nondeterministically.

1. If $\alpha^2 \rightarrow \alpha$ is applied, then σ_1 removes two spikes and emits a spike to σ_2 and σ_3. Thus, neurons σ_2 and σ_3 each receive the spike. Therefore, $C_1 = (0, 1, 1, 0)$.

   At timestep $\tau = 2$, because of one spike in σ_3, $\alpha \rightarrow \alpha$ is used to emit one spike to σ_4. At the same time, due to one spike in σ_2, $\alpha \rightarrow \alpha$ is used to emit a spike to σ_3 and σ_4, respectively. Thus, $C_2 = (0, 0, 1, 2)$.

   At timestep $\tau = 3$, because of two spikes in σ_4, $\alpha^+/\alpha \rightarrow \alpha$ is used to consume one spike and output a spike. At the same time, due to one spike in σ_3, $\alpha \rightarrow \alpha$ is used to emit one spike to neuron σ_4. Thus, $C_3 = (0, 0, 0, 2)$.

   At timestep $\tau = 4$, because of two spikes in σ_4, $\alpha^+/\alpha \rightarrow \alpha$ is used to consume one spike and output a spike. Thus, $C_4 = (0, 0, 0, 1)$.

   At timestep $\tau = 5$, due to a spike in σ_4, $\alpha^+/\alpha \rightarrow \alpha$ is used to consume one spike and output a spike. Thus, $C_5 = (0, 0, 0, 0)$. At this time, the system halts.

2. If $\alpha^+/\alpha \to \alpha$ is used, then σ_1 removes a spike and emits a spike to σ_2 and σ_3. Thus, neurons σ_2 and σ_3 each receive the spike. Therefore, $C_1 = (1, 1, 1, 0)$.
   At timestep $\tau = 2$, due to one spike in σ_3, $\alpha \to \alpha$ is used to emit a spike to σ_4. Due to a spike in neuron σ_2, its rule $\alpha \to \alpha$ is applied to send a spike to neurons σ_3 and σ_4, respectively. At the same time, neuron σ_1 uses $\alpha^+/\alpha \to \alpha$ to send a spike to neurons σ_2 and σ_3. Thus, $C_2 = (0, 1, 2, 2)$.
   At time $\tau = 3$, due to two spikes in neuron σ_4, its rule $\alpha^+/\alpha \to \alpha$ is used to consume one spike and output a spike. Because of two spikes in σ_3, $\alpha^2 \to \lambda$ is used to consume the two spikes. At the same time, σ_2 uses $\alpha \to \alpha$ to emit one spike to σ_3 and σ_4. Thus, $C_3 = (0, 0, 1, 2)$.
   At timestep $\tau = 4$, because of two spikes in σ_4, $\alpha^+/\alpha \to \alpha$ is used to remove one spike and output one spike. At the same time, σ_3 uses $\alpha \to \alpha$ to emit one spike to σ_4. Therefore, $C_4 = (0, 0, 0, 2)$.
   At timestep $\tau = 5$, because of two spikes in σ_4, $\alpha^+/\alpha \to a$ is used to consume one spike and output a spike. Thus, $C_5 = (0, 0, 0, 1)$.
   At timestep $\tau = 6$, because of one spike in σ_4, $\alpha^+/\alpha \to \alpha$ is used to consume one spike and output a spike. Thus, $C_6 = (0, 0, 0, 0)$. At this time, the system stops.

Finally, the SNP system generates the number set $\{3, 4\}$.

2.4 Spiking Neural P Systems with Extended Rules

The work by Ref. [16] presented an enhanced variant of spiking neural P systems, termed either spiking neural P systems with extended rules or extended SNP systems, characterized by the replacement of standard spiking/firing rules with their extended counterparts. In the extended SNP systems, extended rules of the form $F/\alpha^c \to \alpha^p$ are introduced to replace original spiking/firing rules of form $F/\alpha^c \to \alpha$ and forgetting rules of form $\alpha^s \to \lambda$ in SNP systems.

2.4.1 Definition

Definition 2.2 An extended SNP system with n neurons is given by

$$\Pi = (O, \sigma_1, \ldots, \sigma_n, syn, in, out) \tag{2.3}$$

wherein

1. $O = \{\alpha\}$ is the singleton alphabet, where α is called the spike;
2. $\sigma_1, \sigma_2, \ldots, \sigma_n$ are n neurons, with $\sigma_i = (v_i, R_i),\ i = 1, 2, \ldots, n$, wherein
   a. $v_i \geq 0$ is the number of spikes in σ_i;
   b. R_i is the set of rules, $F/\alpha^c \to \alpha^p$, in which F is the regular expression over α, and $c \geq p, c \geq 1, p \geq 0$;

3. $syn \subseteq \{1, \ldots, n\} \times \{1, \ldots, n\}$; for all $(i, j) \in syn$, $i \neq j$, and $1 \leq i, j \leq n$ (synapses);
4. $out, in \in \{1, \ldots, n\}$, respectively, indicate the output neuron and input neuron.

From a topological perspective, extended SNP systems adopt the structure of digraphs, with neurons represented as vertices and synaptic connections depicted as directed edges. The system's dynamic state is quantified by the spike counts within each neuronal unit.

The extended SNP systems have spiking/firing rules $F/\alpha^c \to \alpha^p$. Note that when $p = 0$, the rule can be rewritten by $F/\alpha^c \to \lambda$, known as forgetting rule.

A rule $F/\alpha^c \to \alpha^p$ in neuron σ_i can be used as below. If σ_i has v spikes, and $\alpha^v \in L(F)$, $v \geq c$, then the neuron can fire, and applying the rule means that c spikes are consumed, and p spikes are generated. Then, all its post-synaptic neuronal units shall acquire p spike inputs. All components maintain synchronized operation by adhering to a common clock throughout the system.

Based on the semantics of these rules, neuron σ_i's state evolution can be mathematically described by

$$v_i(\tau) = \begin{cases} v_i(\tau - 1) - c + m, & \text{if neuron } \sigma_i \text{ fires;} \\ v_i(\tau - 1) + m, & \text{otherwise.} \end{cases} \tag{2.4}$$

where $v_i(\tau - 1)$ and $v_i(\tau)$ are the states of σ_i at timesteps $\tau - 1$ and τ, respectively; c is the consumed number of spikes; m is the received number of spikes from its presynaptic neurons.

A rule expressed as $F/\alpha^c \to \alpha^p$ may be simplified to $\alpha^c \to \alpha^p$ if F is identical to α^c.

When rules $F_1/\alpha^{c_1} \to \alpha^{p_1}$ and $F_2/\alpha^{c_2} \to \alpha^{p_2}$ share overlapping languages ($L(F_1) \cap L(F_2) \neq \emptyset$), multiple rules may become applicable within a single neuron. In such scenarios, the system selects one rule arbitrarily through nondeterministic choice.

During the computation, a configuration at timestep τ can be expressed as $C_\tau = (v_1(\tau), v_2(\tau), \ldots, v_n(\tau))$, where $v_i(\tau)$ is the number of spikes in σ_i at timestep τ, i.e., the state of σ_i at timestep τ, $i = 1, 2, \ldots, n$. Therefore, $C_0 = (v_1, v_2, \ldots, v_n)$ denotes initial configuration. A state change occurs when moving between different system configurations. The complete series of such state changes originating from the startup configuration constitutes a computational process. This process terminates upon encountering a configuration that does not permit further rule applications.

In Ref. [16], the extended SNP systems were proven to be Turing universal computing devices, including language generating devices, function computing devices and number accepting/generating devices.

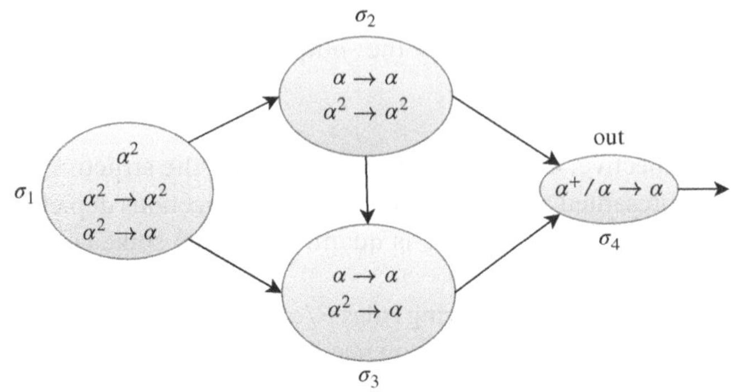

Figure 2.2 An extended SNP system (provided by authors).

2.4.2 Illustrative Example

Figure 2.2 demonstrates the operational principles of the extended SNP system through a basic case study. This system architecture consists of four neuronal units (σ_1 through σ_4), with σ_4 designated as the output neuron.

At the beginning, neuron σ_1 contains two spikes, and neurons σ_2, σ_3 and σ_4 have no spikes. Thus, $C_0 = (2, 0, 0, 0)$.

At timestep $\tau = 1$, due to two spikes in σ_1, $\alpha^2 \to \alpha^2$ and $\alpha^2 \to \alpha$, can be used. Thus, one of them is selected nondeterministically.

1. If $\alpha^2 \to \alpha^2$ is applied, then σ_1 removes two spikes and emits two spikes to σ_2 and σ_3 respectively. Thus, neurons σ_2 and σ_3 each receives the two spikes. Therefore, $C_1 = (0, 2, 2, 0)$.

   At time $\tau = 2$, because of two spikes in σ_3, $\alpha^2 \to \alpha$ is used to emit one spike to neuron σ_4. At the same time, because of two spikes in σ_2, $\alpha^2 \to \alpha^2$ is used to emit two spikes to σ_3 and σ_4, respectively. Thus, $C_2 = (0, 0, 2, 3)$.

   At time $\tau = 3$, because of three spikes in σ_4, $\alpha^+/\alpha \to \alpha$ is used to consume one spike and output a spike. At the same time, because of two spikes in σ_3, $\alpha^2 \to \alpha$ is used to emit one spike to σ_4. Thus, $C_3 = (0, 0, 0, 3)$.

   At timestep $\tau = 4$, because of three spikes in σ_4, $\alpha^+/\alpha \to \alpha$ is used to remove one spike and output a spike. Thus, $C_4 = (0, 0, 0, 2)$.

   At timestep $\tau = 5$, because of two spikes in σ_4, $\alpha^+/\alpha \to \alpha$ is used to consume one spike and output a spike. Thus, $C_4 = (0, 0, 0, 1)$.

   At timestep $\tau = 6$, because of one spike in σ_4, $\alpha^+/\alpha \to \alpha$ is used to consume a spike and output a spike. Thus, $C_5 = (0, 0, 0, 0)$. At this time, the system stops.
2. If $\alpha^2 \to \alpha$ is used, then σ_1 removes two spikes and emits a spike to σ_2 and σ_3. Thus, neurons σ_2 and σ_3 each receives the spike. Therefore, $C_1 = (0, 1, 1, 0)$.

At timestep $\tau = 2$, because of one spike in σ_3, $\alpha \rightarrow \alpha$ is used to emit one spike to σ_4. At the same time, due to one spike in σ_2, $\alpha \rightarrow \alpha$ is applied to emit one spike to neurons σ_3 and σ_4, respectively. Thus, $C_2 = (0, 0, 1, 2)$.

At time $\tau = 3$, due to two spikes in neuron σ_4, its rule $\alpha^+/\alpha \rightarrow a$ is used to consume a spike and output a spike. At the same time, due to one spike in σ_3, $\alpha \rightarrow \alpha$ is used to emit one spike to neuron σ_4. Thus, $C_3 = (0, 0, 0, 2)$.

At timestep $\tau = 4$, because of two spikes in σ_4, $\alpha^+/\alpha \rightarrow a$ is used to remove one spike and output a spike. Thus, $C_4 = (0, 0, 0, 1)$.

At timestep $\tau = 5$, because of one spike in σ_4, $\alpha^+/\alpha \rightarrow \alpha$ is used to consume a spike and output a spike. Thus, $C_6 = (0, 0, 0, 0)$. At this time, the system halts.

Finally, the extended SNP system generates the number set $\{3, 4\}$.

2.5 Spiking Neural P Systems with Autapses

Building upon the autaptic mechanism, Ref. [17] introduced spiking neural P systems with autapses, an extension of spiking neural P systems that incorporates self-connections, termed as SNP-AU systems. Unlike traditional SNP systems, SNP-AU systems integrate an autapse-based structural modification. This autapse mechanism means that each neuron has an output feedback.

2.5.1 Definition

Definition 2.3 A SNP-AU system with n neurons is a tuple

$$\Pi = (O, \sigma_1, \ldots, \sigma_n, syn, in, out) \tag{2.5}$$

wherein

1. $O = \{\alpha\}$ is the singleton alphabet, where α is called the spike;
2. $\sigma_1, \sigma_2, \ldots, \sigma_n$ are n neurons, with $\sigma_i = (v_i, R_i),\ i = 1, 2, \ldots, n$, wherein
   a. $v_i \geq 0$ is the number of spikes in neuron σ_i;
   b. R_i is the set of rules, with $F/\alpha^c \rightarrow \alpha^p$, in which F is the regular expression over α, and $c \geq p$, $p \geq 0$, and $c \geq 1$;
3. $syn \subseteq \{1, \ldots, n\} \times \{1, \ldots, n\} \times K$ indicates the synapses among neurons; Some neurons may possess autaptic connections; Each synaptic connection is represented as a triplet (i, j, k), where i and j range from 1 to n (with $i = j$ indicating autapses), and k belongs to set K (denoting possible synapse counts). Here, k specifies the quantity of synapses from σ_i to σ_j, where interneuronal connections ($i \neq j$) always have exactly one synapse (independent of k), while autaptic connections ($i = j$) contain k synapses;
4. $out, in \in \{1, \ldots, n\}$, respectively, denote the output neuron and input neuron.

This system can be represented as a directed graph, with neuron units serving as vertices and directed edges symbolizing interneuronal synaptic linkages. In contrast to both standard and extended SNP models, the SNP-AU framework represents synaptic connections through triplets, $syn \subseteq \{1, \ldots, n\} \times \{1, \ldots, n\} \times K$. When the indices satisfy $i = j$, parameter k denotes the presence of an autaptic connection in the neuron. Under the simplified condition that each neuron contains no more than one autapse, we assign $k = 1$ for all cases where i equals j.

The neuronal firing pattern follows a specific rule $F/\alpha^c \to \alpha^p$. Neuron σ_i triggers this mechanism if and only if its spike count v reaches or exceeds threshold c, with the additional constraint $\alpha^v \in L(F)$. This rule's execution results in the consumption of c spikes by σ_i, with immediate transmission of p spikes to all postsynaptic neurons σ_j defined by synaptic connections (i, j, k), even when $j = i$ in autaptic cases. For the sake of simplicity, the spiking rule is denoted by $\alpha^c \to \alpha^p$ if $v = c$ and $F = \alpha^c$.

Based on the semantics of these rules, state equation for σ_i is given by

$$v_i(\tau) = \begin{cases} v_i(\tau - 1) - c + m + p, & \text{if neuron } \sigma_i \text{ fires;} \\ v_i(\tau - 1) + m, & \text{otherwise.} \end{cases} \tag{2.6}$$

where $v_i(\tau - 1)$ and $v_i(\tau)$ are the states of σ_i at timesteps $\tau - 1$ and τ, respectively; c is the consumed number of spikes; p is the produced number of spikes; m is the received number of spikes from its presynaptic neurons.

When $p = 0$, this rule is expressed by $F/\alpha^s \to \lambda$, known as the forgetting rule. Applying the forgetting rule means that neuron σ_i removes s spikes, and does not produce any spike.

Similarly, in a neuron, non-determinacy is used for rule selection when two or more rules can be applied. Due to the assumption that the global clock synchronizes all neurons, they run in parallel with each other. Nonetheless, within every neuron, the rules are applied in a consecutive manner, allowing no more than a single rule to be active simultaneously.

Let $C_\tau = (v_1(\tau), v_2(\tau), \ldots, v_n(\tau))$ denote the configuration at time τ, where $v_i(\tau)$ is the state of neuron σ_i at time τ, indicating the number of spikes in neuron σ_i at time τ, $i = 1, 2, \ldots, n$. The starting state may be represented as the tuple $C_0 = (v_1, v_2, \ldots, v_n)$. Configurations evolve through discrete transition steps. From the starting point, any connected series of these transitions constitutes a computation. Such computation stops when reaching a configuration that triggers no applicable rules.

In Ref. [17], SNP-AU systems were validated as computationally universal machines in the Turing sense, capable of performing number generation/acceptance tasks and executing function calculations.

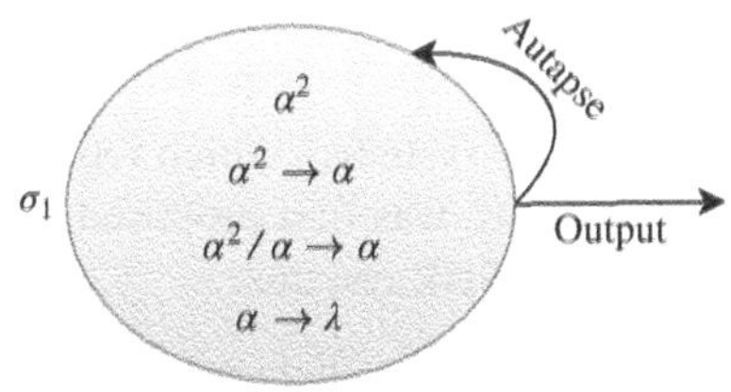

Figure 2.3 A SNP-AU system. Adapted from [17].

2.5.2 Illustrative Example

Figure 2.3 shows a simple example of SNP-AU system to explain the working mechanism. The SNP-AU system contains only one neuron with autapse, σ_1, and it is also an output neuron. Initially, neuron σ_1 contains two spikes.

At timestep $\tau = 1$, due to two spikes in σ_1, $\alpha^2 \to \alpha$ and $\alpha^2/\alpha \to \alpha$ can be used. Thus, one of them is selected nondeterministically.

1. If $\alpha^2 \to \alpha$ is used, then σ_1 consumes two spikes and outputs one spike. Due to autapse, σ_1 receives a spike. Thus, neuron σ_1 contains a spike. At timestep 2, due to a spike in σ_1, $\alpha \to \lambda$ is used to consume the spike. As a result, the system stops.
2. If rule $\alpha^2/\alpha \to \alpha$ is used, then neuron σ_1 consumes only one spike and outputs one spike. Due to autapse, σ_1 receives a spike. Therefore, σ_1 contains two spikes, that is, it returns to its initial state.

   At timestep $\tau = 2$, because of two spikes in σ_1, $\alpha^2 \to \alpha$ and $\alpha^2/\alpha \to \alpha$, can be used. Thus, one of them is selected nondeterministically.
   a. If rule $\alpha^2 \to \alpha$ is used, then σ_1 removes two spikes and outputs a spike. Due to autapse, neuron σ_1 receives a spike. Thus, neuron σ_1 contains a spike. At timestep $\tau = 3$, because of one spike in σ_1, $\alpha \to \lambda$ is used to remove the spike. Thus, the system halts.
   b. If $\alpha^2/\alpha \to \alpha$ is applied, then σ_1 removes a spike and outputs a spike. Due to autapse, neuron σ_1 receives a spike. Therefore, σ_1 contains two spikes, that is, it returns to its initial state again.

From the above process, it can be seen that in the initial state or when the system returns to the initial state, if rule $\alpha^2 \to \alpha$ is executed, the system outputs a spike and stops. However, if the system executes rule $\alpha^2/\alpha \to \alpha$, it will return to its initial state after outputting a spike.

Finally, the number set $\{n \mid n = 1, 2, \ldots\}$ is generated by the system, that is, all nonzero integers.

2.6 Nonlinear Spiking Neural P Systems

The concept of nonlinear spiking neural P systems (NSNP systems), which represent an extended nonlinear form of standard SNP systems, was first proposed in [18]. There are three key distinctions between NSNP and SNP systems:

1. Neurons possess a continuous real-valued state reflecting their membrane potential;
2. The firing mechanism is triggered under the condition $v_i \geq T$, with T being a strictly positive threshold parameter ($T \in \mathbb{R}^+$);
3. Nonlinear spiking rule $T|a^{g(v_i)} \rightarrow a^{f(v_i)}$ is introduced.

The nonlinear spiking rule indicates nonlinear spiking mechanisms, including nonlinear spike generation and nonlinear spike consumption. This is the most fundamental difference between NSNP systems and SNP (and variants). Therefore, NSNP systems have nonlinear processing capabilities and can be used to model nonlinear dynamical systems.

2.6.1 Definition

Definition 2.4 A NSNP system with n neurons is given by

$$\Pi = (O, \sigma_1, \sigma_2, \ldots, \sigma_n, syn, in, out) \tag{2.7}$$

wherein

1. $O = \{a\}$ is the singleton alphabet, where a is called the spike;
2. $\sigma_1, \ldots, \sigma_n$ are n neurons, of the form

   $$\sigma_i = (v_i, R_i), \ 1 \leq i \leq n$$

   wherein
   a. $v_i \in \mathbb{R}^+$ denotes the value of spikes in neuron σ_i, indicating initial state of σ_i;
   b. R_i denotes the set of spiking rules, with $T|a^{g(v_i)} \rightarrow a^{f(v_i)}$, in which $T \in \mathbb{R}^+$ is the threshold for firing, functions $f(v_i)$ and $g(v_i)$ are the nonlinear or linear, and $g(v_i) \geq f(v_i) \geq 0$.
3. $syn \subseteq \{1, 2, \ldots, n\} \times \{1, 2, \ldots, n\}$; for all $(i, j) \in syn$, $i \neq j$, $i, j = 1, 2, \ldots, n$ (synapses).
4. $out, in \in \{1, 2, \ldots, n\}$, respectively, assign output neuron and input neuron.

In standard representations, a NSNP architecture adopts a digraph model wherein neuronal units serve as graph nodes while directional edges symbolize interneuronal synaptic pathways.

In contrast to SNP models, NSNP systems represent neuronal states through positive real-valued quantities (denoted by $\mathbb{R}^+$, the domain of nonnegative real numbers). The neuronal state evolution in NSNP architectures is driven by spike-processing rules of nonlinear nature, mathematically represented by $T|\alpha^{g(v_i)} \rightarrow \alpha^{f(v_i)}$. In this context, $g(v_i)$ serves as the consumption function that operates on neuron σ_i's state v_i through linear or nonlinear relationships, whereas the generation function $f(v_i)$ exclusively employs nonlinear operations on σ_i's state variable v_i. Below are four functions, which can be used as $g(\cdot)$ and $f(\cdot)$:

1. $f_1(v) = \begin{cases} 0, & v \leq 0 \\ 2, & v > 0 \end{cases}$
2. $f_2(v) = \begin{cases} 0, & v \leq 0 \\ v, & v > 0 \end{cases}$
3. Logistic function:
$$f_3(v) = \frac{1}{1 + e^{-cv}} \tag{2.8}$$
4. Sigmoid function:
$$f_4(v) = \tanh(v) = \frac{e^{2v} - 1}{e^{2v} + 1} \tag{2.9}$$

In Ref. [18], $f_1(v)$ and $f_2(v)$ were used to discuss the computational completeness of NSNP systems. In subsequent chapters, $f_3(v)$ and $f_4(v)$ will be used in the recurrent-like models and echo-like models discussed.

Let $v_i(\tau)$ be the state of σ_i at timestep τ. If satisfying $v_i(\tau) \geq T$ and $v_i(\tau) \geq g(v_i(\tau))$, $T|\alpha^{g(v_i)} \rightarrow \alpha^{f(v_i)}$ can be used. When using the rule, then spikes of $g(v_i(\tau))$ are removed, and the spikes of $f(v_i(\tau))$ are produced. The spikes computed as $f(v_i(\tau))$ are sent to the postsynaptic neurons $\{\sigma_j | (i, j) \in syn\}$. When the function $f(v_i(\tau))$ equals zero, the neuronal operation is termed the forgetting rule, denoted as $T|a^{g(v_i)} \rightarrow \lambda$. This rule implementation results in the elimination of spikes valued at $g(v_i(\tau))$ from the neuron without producing any new spike.

Following the operational principles of spiking and forgetting rules previously defined, the neuronal state equation for σ_i is formulated as:

$$v_i(\tau + 1) = \begin{cases} v_i(\tau) - g(v_i(\tau)) + y_i(\tau), & \text{if } \sigma_i \text{ fires} \\ v_i(\tau) + y_i(\tau), & \text{otherwise} \end{cases} \tag{2.10}$$

wherein at consecutive timesteps τ and $\tau + 1$, neuron σ_i exhibits potential values $v_i(\tau)$ and $v_i(\tau + 1)$ respectively. The term $g(v_i(\tau))$ denotes spikes consumed by neuronal rules, whereas $y_i(\tau)$ captures synaptic inputs transmitted to σ_i. Given that both $g(v_i(\tau))$ and $y_i(\tau)$ may exhibit linear or nonlinear dependence on the neuronal state $v_i(\tau)$ in σ_i, the state equation (2.10) consequently manifests as a nonlinear difference equation.

Because $T_1|\alpha^{g_1(v_i)} \rightarrow \alpha^{f_1(v_i)}$ and $T_2|\alpha^{g_2(v_i)} \rightarrow \alpha^{f_2(v_i)}$ in neuron σ_i may meet $\{v_i \geq T_1\} \cap \{v_i \geq T_2\} \neq \emptyset$. Thus, two distinct situations are examined:

1. If $T_1 \neq T_2$, maximum threshold strategy is used. For instance, if $T_1 > T_2$, then rule $T_1|\alpha^{g_1(v_i)} \rightarrow \alpha^{f_1(v_i)}$ is selected to apply, and $T_2|\alpha^{g_2(v_i)} \rightarrow \alpha^{f_2(v_i)}$ can not be used. It should be noted that one of them may be a forgetting rule. For instance, if $T_1|\alpha^{g_1(v_i)} \rightarrow \lambda$ and $T_2|\alpha^{g_2(v_i)} \rightarrow \alpha^{f_2(v_i)}$ can be applied, rule $T_1|\alpha^{g_1(v_i)} \rightarrow \lambda$ is applied but rule $T_2|\alpha^{g_2(v_i)} \rightarrow a^{f_2(v_i)}$ is not used.
2. If $T_1 = T_2 = T$, nondeterministic rule selection strategy is adopted. If $T|\alpha^{g_1(v_i)} \rightarrow \alpha^{f_1(v_i)}$ and rule $T|\alpha^{g_2(v_i)} \rightarrow \alpha^{f_2(v_i)}$ can be used, one of them is selected non-deterministically.

Consistent with standard SNP architectures, neuronal rules execute sequentially within individual processing units while maintaining system-wide parallelism. Consequently, NSNP architectures exhibit inherently distributed, parallel and nondeterministic computational characteristics.

Each neural unit's operational state collectively determines the entire system's configuration at time point τ, $C(\tau) = v_1(\tau), v_2(\tau), \ldots, v_n(\tau))$. Therefore, the starting configuration is represented by $C(0) = (v_1(0), v_2(0), \ldots, v_n(0)) = (v_1, v_2, \ldots, v_n)$. Using forgetting and spiking rules enables the definition of state transitions between configurations. A computational process consists of any valid transition sequence originating from the initial configuration. Termination occurs when the system enters a configuration that permits no further rule applications.

2.6.2 Illustrative Example

The NSNP system shown in Figure 2.4 has two neurons. Initially, neuron σ_1 contains the spike value of 2.5, while neuron σ_2 has no spike. neuron σ_2 is an output neuron. Therefore, the initial configuration is $C_0 = (2.5, 0)$. The $a(v)$ is given by $\lfloor v \rfloor$, that is, $a(v) = \lfloor v \rfloor$.

At timestep $\tau = 1$, two rules of σ_1 can be used. According to maximum threshold strategy, $2|\alpha^v \rightarrow \alpha^{a(v)}$ is used to emit two spikes to σ_2.

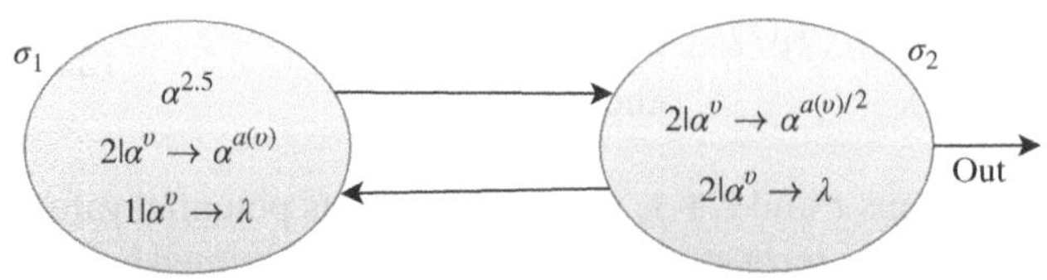

Figure 2.4 A NSNP system. Adapted from [18].

At timestep $\tau = 2$, σ_2 has two spikes. Thus, $2|\alpha^v \to \alpha^{a(v)/2}$ and $2|\alpha^v \to \lambda$ can be used. Due to the same threshold in the two rules, one of them is selected nondeterministically.

1. If rule $2|\alpha^v \to \alpha^{a(v)/2}$ in σ_2 is used, then a spike is emitted to σ_1 and one spike is output.

   At timestep $\tau = 3$, because of one spike in σ_1, $1|\alpha^v \to \lambda$ is used to consume the spike. As a result, the system stops.
2. If $2|\alpha^v \to \lambda$ is used, then it removes two spikes, and the system stops.

The example reveals two distinct processing patterns: a maximum threshold strategy is employed when $\tau = 1$, while a nondeterministic selection mechanism operates at $\tau = 2$.

References

1 M. Ionescu, Gh. Păun, and T. Yokomori. Spiking neural P systems. *Fundamenta Informaticae*, 71: 279–308, 2006.

2 H. Peng and J. Wang. *Advanced Spiking Neural P Systems: Models and Applications*. Springer, Singapore, 2024.

3 Gh. Păun. Spiking neural P systems with astrocyte-like control. *Journal of Universal Computer Science*, 13(11): 1707–1721, 2007.

4 L. Pan and Gh. Păun. Spiking neural P systems with anti-spikes. *International Journal of Computers, Communications & Control*, 4(3): 273–282, 2009.

5 T. Song, L. Pan, and Gh. Păun. Spiking neural P systems with rules on synapses. *Theoretical Computer Science*, 529: 82–95, 2014.

6 T. Wu, A. Păun, Z. Zhang, and L. Pan. Spiking neural P systems with polarizations. *IEEE Transactions on Neural Networks and Learning Systems*, 29(8): 3349–3360, 2017.

7 F.G.C. Cabarle, X. Zeng, N. Murph, T. Son, A. Rodríguez-Patón, and X. Liu. Neural-like P systems with plasmids. *Information and Computation*, 281(4): 104766, 2021.

8 L. Pan, Gh. Păun, G. Zhang, and F. Neri. Spiking neural P systems with communication on request. *International Journal of Neural Systems*, 28(8): 1750042, 2017.

9 H. Peng, B. Li, J. Wang, X. Song, T. Wang, L. Valencia-Cabrera, I. Pérez-Hurtado, A. Riscos-Núñez, and M.J. Pérez-Jiménez. Spiking neural P systems with inhibitory rules. *Knowledge-Based Systems*, 30(2): 2050008, 2020.

10 H. Peng, J. Yang, J. Wang, T. Wang, Z. Sun, X. Song, X. Luo, and X. Huang. Spiking neural P systems with multiple channels. *Neural Network*, 95: 66–71, 2017.

11 X. Liu, X. Tian, Q. Ren, and Y. Zhao. Spiking neural P systems with enzymes. *IEEE Transactions on Nanobioscience*, 21(4): 575–587, 2022.

12 H. Peng, J. Wang, M.J. Pérenz-Jiménez, and A. Riscos-Núñez. Dynamic threshold neural P systems. *Knowledge-Based Systems*, 163: 875–884, 2019.

13 H. Peng and J. Wang. Coupled neural P systems. *IEEE Transactions on Neural Networks and Learning Systems*, 30(6): 1672–1682, 2019.

14 X. Song, L. Valencia-Cabrera, H. Peng, J. Wang, and M.J. Pérez-Jiménez. Spiking neural P systems with delay on synapses. *International Journal of Neural Systems*, 31(1): 2050042, 2021.

15 H. Peng, T. Bao, X. Luo, J. Wang, X. Song, A. Riscos-Núñez, and M.J. Pérez-Jiménez. Dendrite P systems. *Neural Networks*, 127: 110–120, 2020.

16 H. Chen, M. Ionescu, T.O. Ishdorj, A. Păun, Gh. Păun, and M.J. Pérez-Jiménez. Spiking neural P systems with extended rules: universality and languages. *Natural Computing*, 7: 147–166, 2008.

17 X. Song, L. Valencia-Cabrera, H. Peng, and J. Wang. Spiking neural P systems with autapses. *Information Sciences*, 570: 383–402, 2021.

18 H. Peng, Z. Lv, B. Li, X. Luo, J. Wang, X. Song, T. Wang, M.J. Pérez-Jiménez, and A. Riscos-Núñez. Nonlinear spiking neural P systems. *International Journal of Neural Systems*, 30(10): 2050008, 2020.

19 G. Rozenberg and A. Salomaa. *Handbook of Formal Languages, 3 volumes*. Springer-Verlag, Berlin, 1997.

20 A. Salomaa. *Formal Languages*. Academic Press, New York, 1973.

21 Gh. Păun, G. Rozenberg, and A. Salomaa. *The Oxford Handbook of Membrane Computing*. Oxford University Press, New York, 2010.

3

Recurrent-like Models for Time-series Forecasting

3.1 Introduction

With swift progress of the Internet of Things, financial technology, and intelligent manufacturing, time series data has become a core research object in fields such as energy management (such as power load forecasting), financial analysis (such as stock price forecasting), and industrial monitoring (such as equipment failure warning). This type of data has the characteristics of temporal dependence, non-stationarity, and noise sensitivity. Conventional statistical approaches like ARIMA and exponential smoothing demonstrate constrained capabilities when dealing with intricate nonlinear patterns. While ML techniques including SVM and random forests can identify certain nonlinear characteristics, they require manual feature extraction and struggle with modeling distant temporal correlations.

To overcome the bottleneck of traditional methods, recurrent neural networks (RNNs) have become a research hotspot due to their temporal memory ability. RNNs can theoretically model sequences of any length by transmitting historical information through hidden states. However, early RNNs had gradient vanishing/exploding problems, which limited their practical applications. To address the problems of RNN, Ref. [1] presented the long short-term memory (LSTM) model, which achieves long-term preservation of key information through gating mechanisms (output gate, forget gate, and input gate). Moreover, gated recurrent unit (GRU) was proposed to further simplify the structure of LSTM, achieving similar performance with fewer parameters [2, 3]. GRU and LSTM were widely utilized in applications such as energy management, financial analysis, traffic management, medical diagnosis, and industrial monitoring.

In recent years, the cross-fusion of RNN with other technologies has become a new trend. For example, the CNN+RNN model, the RNN model with enhanced attention mechanisms, the hybrid model of RNN and Transformer, the integration

Spiking Neural P Systems for Time Series Analysis, First Edition. Jun Wang and Hong Peng.

of GNN and RNN, knowledge distillation and model pruning technology promote the application of RNN in edge computing devices.

Section 2.6 discusses a nonlinear variant of spiking neural P (SNP) systems, nonlinear spiking neural P (NSNP) systems. NSNP systems are neural-like computational models with nonlinear processing capability, where the nonlinear spiking mechanism is an essential characteristic that distinguishes NSNP systems from other neural-like computational models. The nonlinear spiking mechanism consists of nonlinear spike generation and consumption. In this chapter, the nonlinear spiking mechanism will be used to abstract three recurrent-like models: LSTM-SNP model [4], GSNP model [5], and NSNP-AU model [6]. The NSNP systems in Section 2.6 are theoretical computational models. Therefore, three application variants are first defined, and then the three recurrent-like models are constructed respectively based on them.

3.2 Recurrent Neural Networks

RNNs represent a specialized type of neural architecture characterized by temporal memory retention. These networks feature cyclic connections where neuron outputs can recursively serve as inputs, creating feedback loops within the network topology. RNNs utilize the backpropagation through time (BPTT) method for parameter optimization, wherein error gradients are sequentially calculated backward along the temporal axis. The processing of lengthy sequential data in neural networks frequently leads to gradient-related difficulties (either excessive amplification or diminishment), collectively termed the distant relationship challenge. Advanced recurrent models combat these issues through innovative gated structures, particularly exemplified by LSTM units and their simplified GRU counterparts.

3.2.1 Long Short-term Memory

LSTM models as a specialized type of recurrent neural network, incorporate memory cells and gate structures to overcome RNN limitations [1, 2]. LSTM can effectively address the gradient vanishing or exploding problem of traditional recurrent neural networks. The LSTM architecture utilizes three gating mechanisms for controlling data propagation: the input ($i(\tau)$), forget ($f(\tau)$), and output ($o(\tau)$) gates.

These gates are defined by

1. Input gate: It controls how much information in the current internal state. $\tilde{c}(\tau)$ needs to be saved.

$$i(\tau) = \sigma(W_i x(\tau) + V_i h(\tau - 1) + b_i) \tag{3.1}$$

2. Forget gate: Its function is to modulate the forgetting rate of earlier hidden state values $c(\tau-1)$.

$$f(\tau) = \sigma(V_f h(\tau-1) + W_f x(\tau) + b_f) \tag{3.2}$$

3. Output gate: Its function is to modulate the flow of information from $c(\tau)$ (the internal state) into $h(\tau)$ (the hidden state).

$$o(\tau) = \sigma(V_o h(\tau-1) + W_o x(\tau) + b_o) \tag{3.3}$$

where $\sigma(\cdot)$ is sigmoid function with an output interval of $[0, 1]$, $x(\tau)$ is the current input, and $h(\tau-1)$ is the previous hidden state.

In the recurrent unit in LSTM, the update of internal state and its output are computed by

$$\tilde{c}(\tau) = \tanh(W_c x(\tau) + V_c h(\tau-1) + b_c) \tag{3.4}$$

$$c(\tau) = \tilde{c}(\tau) \odot i(\tau) + c(\tau-1) \odot f(\tau) \tag{3.5}$$

$$h(\tau) = \tanh(c(\tau)) \odot o(\tau) \tag{3.6}$$

In Eqs. (3.1)–(3.4), W_i, V_i, W_f, V_f, W_o, V_o, W_c, and V_c are the matrices, and b_c, b_o, b_f, and b_i are bias vectors. These parameter matrices and bias vectors are trained using the BPTT algorithm.

The architecture of the LSTM's recurrent unit is illustrated in Figure 3.1. The computational workflow proceeds through these sequential steps: (i) Initial calculations derive three gating mechanisms and a candidate state $\tilde{c}(\tau)$ using both the preceding hidden state $h(\tau-1)$ and present input $x(\tau)$; (ii) Memory cell $\mathbf{c}_t$ gets

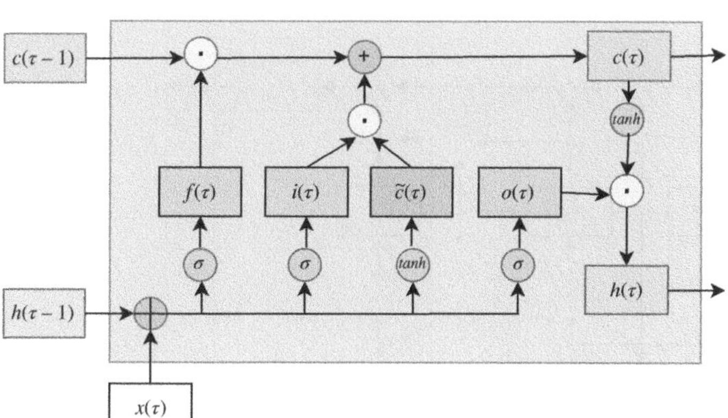

Figure 3.1 The structure of the recurrent unit in LSTM (provided by authors).

modified through the coordinated action of the forgetting mechanism $f(\tau)$ and input regulation $i(\tau)$; (iii) Through integration with the output control $o(\tau)$, the internal state data propagates to become the new hidden state $h(\tau)$.

3.2.2 Gated Recurrent Unit

Compared with LSTM networks, GRUs [3] represent a computationally efficient RNN variant with simplified gate mechanisms. Unlike LSTM, GRU alters the information flow mechanism by employing an update gate, which dynamically balances the retention of historical state data and the integration of new candidate state inputs.

GRU has two gates: reset gate $r(\tau)$ and update gate $z(\tau)$. The two gates are given by

$$r(\tau) = \sigma(W_r x(\tau) + V_r h(\tau - 1) + b_r) \tag{3.7}$$

$$z(\tau) = \sigma(W_z x(\tau) + V_z h(\tau - 1) + b_z) \tag{3.8}$$

The candidate state and state update are given by

$$\tilde{h}(\tau) = \tanh(V_h(h(\tau - 1) \odot r(\tau)) + W_h x(\tau) + b_h) \tag{3.9}$$

$$h(\tau) = (1 - z(\tau)) \odot \tilde{h}(\tau) + z(\tau) \odot h(\tau - 1) \tag{3.10}$$

In Eqs. (3.7)–(3.9), V_h, W_h, V_z, W_z, V_r, and W_r are parameter matrices, and b_h, b_z, and b_r are bias vectors. These parameter matrices and bias vectors are trained using the BPTT algorithm.

Figure 3.2 describes the recurrent unit structure of GRU. The computation process is explained as follows: (i) two gates $r(\tau)$ and $z(\tau)$ are calculated based on

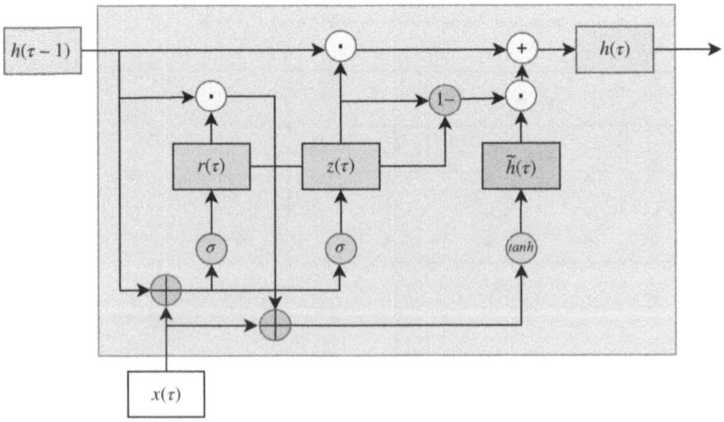

Figure 3.2 The structure of the recurrent unit in GRU (provided by authors).

the previous hidden state $h(\tau-1)$ and current input $x(\tau)$; (ii) the candidate hidden state $\tilde{h}(\tau)$ is calculated using a reset gate $r(\tau)$; (iii) the update gate is utilized to combine previous hidden states and candidate hidden states for updating the current hidden state.

3.3 LSTM-SNP Model

3.3.1 Type-I Nonlinear Gated Spiking Neural P Systems

In Definition 2.4, nonlinear variant of spiking neural P systems is introduced, called NSNP systems. The NSNP systems are theoretical computational models, and used in research on computational completeness. To process sequential data, NSNP systems will be extended to nonlinear gated variants. In this section, the first gated variant of nonlinear spiking neural P systems is discussed, termed type-I nonlinear gated spiking neural P systems (in short, NGSNP-I systems). The variant is first defined, and then its working mechanism is described.

Definition 3.1 A NGSNP-I system is a tuple:

$$\Pi = (O, \sigma_1, \sigma_2, \ldots, \sigma_n, syn)$$

where

1. $O = \{a\}$ is the singleton alphabet, in which a is known as the spike;
2. $\sigma_1, \sigma_2, \ldots, \sigma_n$ are n neurons, of the form $\sigma_i = (v_i, R_i, (r_i, c_i, o_i)),\ i = 1, 2, \ldots, n$, where
   a. $v_i \in \mathbb{R}$ is the state of σ_i, wherein $\mathbb{R}$ indicates the real number set;
   b. R_i is the set of spiking rules, $a^{g(v)} \to a^{f(v)}$, wherein functions $g(\cdot)$ and $f(\cdot)$ are linear or nonlinear;
   c. $r_i = \rho_{i1}(x_i, v_i)$, $c_i = \rho_{i2}(x_i, v_i)$ and $o_i = \rho_{i3}(x_i, v_i)$ are three nonlinear gates, where $x_i \in \mathbb{R}$ is the external input;
3. $syn \subseteq \{1, 2, \ldots, n\} \times \{1, 2, \ldots, n\}$, for all $(i, j) \in syn$, $i \neq j$, $i, j = 1, 2, \ldots, n$ (synapses).

The NGSNP-I system introduces three modifications when contrasted with the NSNP frameworks described in Section 2.6.

1. The stats of each neuron is changed from $v_i \in \mathbb{R}^+$ to $v_i \in \mathbb{R}$.
2. The firing threshold T is removed, thus, each neuron can always fire.
3. Some constraints on $f(\cdot)$ and $g(\cdot)$ are removed.

A NGSNP-I system is expressed by a directed graph of neurons, in which the arcs indicate the synapses between these neurons. Each neuron has an internal state unit, and the state of neuron σ_i at time τ is denoted by $v_i(\tau)$. Additionally, every neuron has nonlinear spiking rule $a^{g(v)} \to a^{f(v)}$, in which functions $g(\cdot)$ and $f(\cdot)$

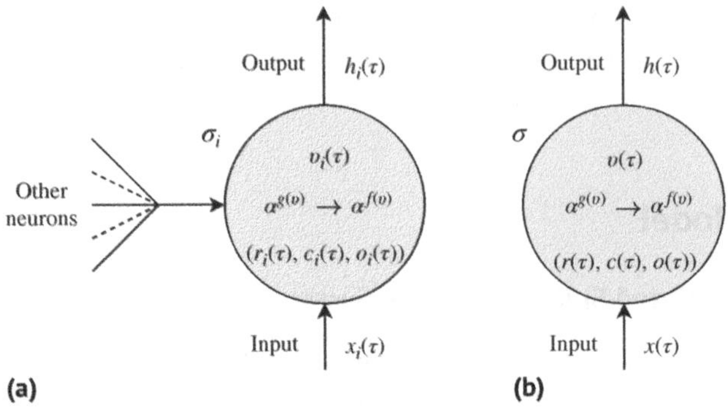

Figure 3.3 (a) In a NGSNP-I system, neuron σ_i and its presynaptic neurons are shown; (b) only one neuron σ is designed in a specialized NGSNP-I system [4]/with permission of Elsevier.

are linear or nonlinear, called spike consumption function and spike generation function, respectively.

Figure 3.3a shows a nonlinear spiking neuron σ_i and its presynaptic in the NGSNP-I system. When both $f(\cdot)$ and $g(\cdot)$ are linear, the resulting spiking rule follows a conventional form. Although $f(\cdot)$ and $g(\cdot)$ may differ, they can also be identical. Compared with the NSNP systems, the difference of the NGSNP-I system is that it has nonlinear gates: $r_i = \rho_{i1}(x_i, v_i)$, $c_i = \rho_{i2}(x_i, v_i)$ and $o_i = \rho_{i3}(x_i, v_i)$. These gating mechanisms serve as regulatory components to achieve recurrent network behavior, specifically governing neuronal state transitions, spike depletion processes, and action potential initiation.

The rule $\alpha^{g(v)} \rightarrow \alpha^{f(v)}$ has two mechanisms: (i) spike generation and (ii) spike consumption. This indicates the nonlinear spiking mechanism of NGSNP-I system. Upon firing, neuron σ_i consumes spikes valued at $g(v)$, produces new spikes with value $f(v)$, and transmits them to its postsynaptic neurons. Thus, $f(\cdot)$ is termed spike generation function, and $g(\cdot)$ is termed spike consumption function. Based on the nonlinear spiking mechanism, the state equation for neuron σ_i is expressed by

$$v_i(\tau) = r_i(\tau) \cdot v_i(\tau - 1) - c_i(\tau) \cdot g(x_i(\tau) + v_i(\tau - 1)) + m \tag{3.11}$$

where $r_i(\tau) = \rho_{i1}(x_i(\tau) + v_i(\tau - 1))$, $c_i(\tau) = \rho_{i2}(x_i(\tau) + v_i(\tau - 1))$; $v_i(\tau)$ and $v_i(\tau - 1)$ are the states of neuron σ_i at time τ and time $\tau - 1$, respectively; $x_i(\tau)$ is an external input of neuron σ_i at time τ, and m is the value of spikes received from its presynaptic neurons; $\rho_{i1}(\cdot)$ and $\rho_{i2}(\cdot)$ are two nonlinear functions. Moreover, neuron σ_i outputs the spikes:

$$h_i(\tau) = o_i(\tau) \cdot f(x_i(\tau) + v_i(\tau - 1)) \tag{3.12}$$

wherein $h_i(\tau)$ is the output, and $o_i(\tau) = \rho_{i3}(x_i(\tau) + v_i(\tau - 1))$.

The NGSNP-I system functions through an iterative computational process. Beginning from predefined starting states, the system operates in a continuous cycle until meeting its termination criteria. Neural states are dynamically modified according to Eq. (3.11), while corresponding outputs are produced through the application of Eq. (3.12).

3.3.2 LSTM-SNP

The NGSNP-I framework is adapted to develop a novel recurrent-type architecture, where the modified system operates under two key constraints: (a) a single-neuron configuration is adopted; (b) the generation and consumption functions are identical ($g(\cdot) \equiv f(\cdot)$).

Figure 3.3b shows the neuron σ in the specialized NGSNP-I system. Due to only one neuron σ, $m = 0$ in Eq. (3.11). The dynamical behavior of neuron σ is characterized by its state and output equations:

$$v(\tau) = r(\tau) \cdot v(\tau - 1) - c(\tau) \cdot f(v(\tau - 1) + x(\tau)) \tag{3.13}$$

$$h(\tau) = o(\tau) \cdot f(x(\tau) + v(\tau - 1)) \tag{3.14}$$

where $v(\tau)$ is the state of neuron σ at time τ, and $v(\tau - 1)$ is the state of neuron σ at time $\tau - 1$; $h(\tau)$ is the output of neuron σ at timestep τ, and $x(\tau)$ is the external input of neuron σ at timestep τ. Three nonlinear gates, $o(\tau)$, $c(\tau)$, and $r(\tau)$, respectively, are called generation, consumption, and reset gates: $r(\tau) = \rho_1(v(\tau - 1) + x(\tau))$, $c(\tau) = \rho_2(v(\tau - 1) + x(\tau))$ and $o(\tau) = \rho_3(v(\tau - 1) + x(\tau))$, wherein $\rho_1(\cdot)$, $\rho_2(\cdot)$, and $\rho_3(\cdot)$ are three nonlinear functions.

Thus, neuron σ becomes a recurrent-like neuron, called LSTM-SNP neuron. The NGSNP-I system in its current form lacks trainable parameters, making it non-trainable. We address this by introducing a parameterized variant. For neuron σ, both the internal state evolution and external response equations are updated accordingly:

$$\begin{cases} v(\tau) = r(\tau) \cdot v(\tau - 1) - c(\tau) f(w_1 \cdot x(\tau) + w_2 \cdot v(\tau - 1) + b) \\ h(\tau) = o(\tau) \cdot f(w_1' \cdot x(\tau) + w_2' \cdot v(\tau - 1) + b') \end{cases} \tag{3.15}$$

where w_1, w_2, w_1', w_2', b, and b' are its trainable parameters. The main purpose is to apply the parameterized NGSNP-I system to construct a learning model.

To achieve broader applicability, the parameterized NGSNP-I framework can be generalized to accommodate multidimensional vectors $\mathbf{x}(\tau)$ (input), $\mathbf{h}(\tau)$ (output),

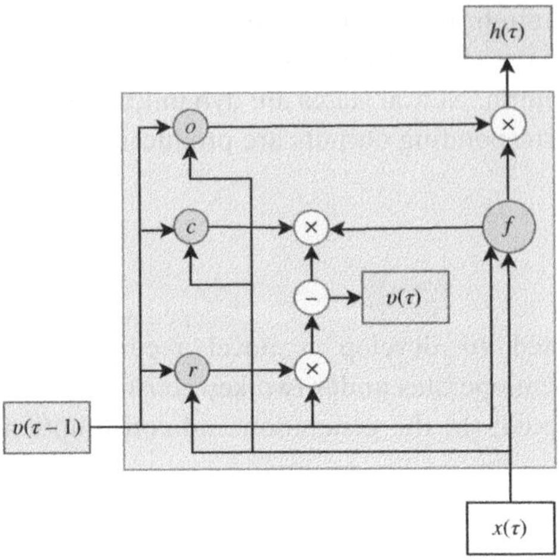

Figure 3.4 Functional structure of LSTM-SNP model [4]/with permission of Elsevier.

and $\mathbf{v}(\tau)$ (state). Consequently, state equation and input–output relationship are reformulated by

$$\begin{cases} \mathbf{v}(\tau) = r(\tau) \cdot \mathbf{v}(\tau - 1) - c(\tau) \cdot f(\mathbf{w}_v \cdot \mathbf{v}(\tau - 1) + \mathbf{w}_x \cdot \mathbf{x}(\tau) + \mathbf{b}) \\ \mathbf{h}(\tau) = o(\tau) \cdot f(\mathbf{w}'_v \cdot \mathbf{v}(\tau - 1) + \mathbf{w}'_x \cdot \mathbf{x}(\tau) + \mathbf{b}') \end{cases} \tag{3.16}$$

where $\mathbf{b}$ and $\mathbf{b}'$ are the biases, and $\mathbf{w}_x$, $\mathbf{w}_v$, $\mathbf{w}'_v$, and $\mathbf{w}'_x$ are the matrices.

Building upon the parameterized NGSNP-I framework, we develop a LSTM architecture variant termed LSTM-SNP (see Ref. [4]). The schematic diagram in Figure 3.4 shows the operational framework of this model. The LSTM-SNP incorporates three distinct nonlinear gating mechanisms to regulate its recurrent operations: the "generation gate," "consumption gate," and "reset gate." These gates are given by

1. Reset gate: It calculates the degree to which prior state information is overwritten, based on the current input, historical state data, and bias.

$$r(\tau) = \rho(\mathbf{V}_r \cdot \mathbf{v}(\tau - 1) + \mathbf{W}_r \cdot \mathbf{x}(\tau) + \mathbf{b}_r) \tag{3.17}$$

2. Consumption gate: The proportion of prior state utilization is calculated based on current input, historical state, and bias parameters.

$$c(\tau) = \rho(\mathbf{W}_c \cdot \mathbf{x}(\tau) + \mathbf{V}_c \cdot \mathbf{v}(\tau - 1) + \mathbf{b}_c) \tag{3.18}$$

3. Generation gate: The quantity of emitted spikes is regulated based on present input signals, historical states, and inherent bias parameters.

$$o(\tau) = \rho(\mathbf{V}_o \cdot \mathbf{v}(\tau - 1) + \mathbf{W}_o \cdot \mathbf{x}(\tau) + \mathbf{b}_o) \tag{3.19}$$

The spikes produced by σ are determined by

$$\mathbf{a}(\tau) = f(\mathbf{V}_a \cdot \mathbf{v}(\tau-1) + \mathbf{W}_a \cdot \mathbf{x}(\tau) + \mathbf{b}_a) \tag{3.20}$$

In Eqs. (3.17)–(3.20), $\mathbf{W}_a$, $\mathbf{V}_a$, $\mathbf{b}_a$, $\mathbf{W}_o$, $\mathbf{V}_o$, $\mathbf{b}_o$, $\mathbf{W}_c$, $\mathbf{V}_c$, $\mathbf{b}_c$, $\mathbf{W}_r$, $\mathbf{V}_r$, and $\mathbf{b}_r$ are the trainable parameters.

Through three nonlinear gates and their resulting spike signals, neuron σ's state and output at timestep τ are determined by

$$\begin{cases} \mathbf{v}(\tau) = r(\tau) \odot \mathbf{v}(\tau-1) - c(\tau) \odot \mathbf{a}(\tau) \\ \mathbf{h}(\tau) = o(\tau) \odot \mathbf{a}(\tau), \end{cases} \tag{3.21}$$

where "$\odot$" denotes the inner product of two vectors.

Figure 3.5 shows the unfolding of the neuron depicted in Figure 3.4 at time $\tau = 1, 2, \ldots, T$. The unfolding structure can be regarded as a RNN. Consequently, the LSTM-SNP architecture is compatible with RNN-based learning paradigms, featuring shared parameter optimization. Similar to standard LSTM networks, this model can be effectively trained using the BPTT method.

Figure 3.5 demonstrates that LSTM-SNP architecture exhibits both short-term memory (capturing the most recent T time steps) and long-term memory capabilities, as evidenced by its optimized parameters. This dual-memory characteristic confirms LSTM-SNP as a specialized LSTM derivative. The LSTM-SNP architecture exhibits three key distinctions from conventional LSTM models: (i) Its recurrent structure draws inspiration from the nonlinear spike processing in NSNP systems; (ii) Distinct mathematical formulations govern both state updating and input–output relationships; (iii) Unlike standard LSTM units that maintain only hidden states, LSTM-SNP neurons possess dual components: internal states and outputs.

When analyzed through membrane computing principles, the LSTM-SNP architecture is fundamentally composed of a single nonlinear neuron with adjustable parameters, as illustrated in Figure 3.4. Conversely, from a recurrent neural network standpoint, this model represents a modified LSTM configuration with

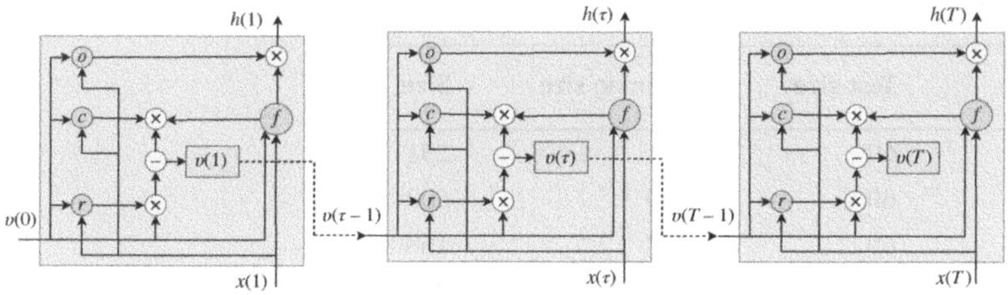

Figure 3.5 LSTM-SNP model for time series forecasting [4]/with permission of Elsevier.

the unfolded architecture depicted in Figure 3.5, capable of undergoing training processes.

3.3.3 Model Evaluation

3.3.3.1 Dataset

The LSTM-SNP model's performance is assessed through a time series prediction task, employing five real-world datasets for experimental benchmarking. These datasets include Milk (monthly milk production per cow), Lake (monthly Lake Erie levels), Closing (monthly closings of the Dow-Jones industrial index), SP500 (S&P 500 stock index), and MG (Mackey–Glass data).

The time series are comprehensively tabulated in Table 3.1, where all feature values have been normalized to the interval between -1 and 1.

3.3.3.2 Evaluation Indicator

In benchmarking tests, the LSTM-SNP architecture is quantitatively compared with leading prediction techniques and fundamental models using root mean square error (RMSE) as the standardized measure. The RMSE is given by

$$\text{RMSE} = \sqrt{\frac{1}{T}\sum_{\tau=1}^{T}(\widetilde{x}(\tau) - x(\tau))^2}, \tag{3.22}$$

where the time series length is denoted by T, $x(\tau)$ represents the observed data and $\widetilde{x}(\tau)$ corresponds to the predicted values. A lower RMSE typically signifies superior predictive accuracy.

3.3.3.3 Experimental Results

Seven state-of-the-art prediction models and five baseline prediction models are used to demonstrate the prediction performance of the LSTM-SNP model. The five baseline prediction models are WAR [7], MAR [8], ANN [9], TCN [10],

Table 3.1 Statistical information of datasets [4]/with permission of Elsevier.

Name	Test size	Training size	Size
Closing	60	231	291
Lake	60	540	600
Milk	60	108	168
SP500	60	191	251
MG	60	940	1,000

and ANFIS [11]. WAR adopts wavelet transforms in its autoregressive framework, contrasting with MAR's multiscale resolution methodology. Prediction systems diverge between ANN's neural network foundation and TCN's hybrid convolutional-recurrent architecture. ANFIS stands distinct as a fuzzy inference system built upon adaptive neural networks.

These state-of-the-art predictive methods include Wavelet-HFCM [12], EMD-HFCM [13], CNN-FCM [14], EWT-ESN [15], ETS-ANN [16], SLD-FIG [17], and KM-HFCM [18]. CNN-FCM represents a hybrid architecture blending CNN and FCM methodologies. The high-order FCM family comprises three specialized models: Wavelet-HFCM applying wavelet analysis, EMD-HFCM using empirical mode decomposition, and KM-HFCM implementing kernel mapping. EWT-ESN combines empirical wavelet processing with reservoir computing, whereas ETS-ANN fuses neural networks with state space smoothing techniques. Distinctively, SLD-FIG implements sequential linear partitioning with adaptive fuzzy granular systems.

The RMSE-based performance evaluation of five time series is displayed in Table 3.2, where the LSTM-SNP model is benchmarked against seven advanced and five basic prediction approaches. The outcomes of the benchmarked approaches are sourced from their respective original publications. Note that in Table 3.2, "–" indicates that the corresponding result was not reported in the original study. The comparative analysis in Table 3.2 indicates: (i) MG series predictions show negligible difference between LSTM-SNP and Wavelet-HFCM,

Table 3.2 The comparisons of LSTM-SNP with other methods in terms of RMSE [4]/with permission of Elsevier.

Model	Closing	Lake	Milk	SP500	MG
WAR	29.8220	0.6380	57.7170	17.7970	0.0350
MAR	26.7330	0.3900	37.8380	16.0410	0.0020
ANN	28.5320	0.4020	27.1130	17.6960	0.0050
TCN	25.2143	0.4093	33.8579	51.2674	0.0008
ANFIS	27.5620	0.0010	9.5780	14.9350	0.4580
CNN-FCM	25.1895	0.3913	30.4738	20.8158	0.0013
Wavelet-HFCM	23.159	0.377	8.258	16.105	0.004
EMD-HFCM	18.875	0.436	7.403	7.139	0.009
EWT-ESN	–	0.3616	12.4524	–	–
ETS-ANN	–	0.3666	6.4125	–	–
SLD-FIG	–	–	27.5078	–	0.056
KM-HFCM	250.076	–	–	12.006	–
LSTM-SNP	24.4279	0.4870	48.3005	11.1704	0.0042

both surpassing competitors. (ii) SP500 forecasting results position EMD-HFCM as the leader, with LSTM-SNP securing runner-up status. (iii) Closing series evaluation highlights the advantage of EMD-HFCM and Wavelet-HFCM over other approaches, leaving LSTM-SNP in third place. (iv) Specialized performance emerges with ETS-ANN excelling in Milk series and EWT-ESN dominating Lake series predictions.

3.4 GSNP Model

3.4.1 Type-II Nonlinear Gated Spiking Neural P Systems

In this section, the second gated variant of nonlinear spiking neural P systems is discussed, called type-II nonlinear gated spiking neural P system (in short, NGSNP-II systems). Unlike NGSNP-I systems, NGSNP-II systems only have two nonlinear gates.

Definition 3.2 A NGSNP-II system is a tuple:

$$\Pi = (O, \sigma_1, \sigma_2, \ldots, \sigma_n, syn)$$

wherein

1. $O = \{\alpha\}$ is the singleton alphabet, in which α is known as the spike;
2. $\sigma_1, \sigma_2, \ldots, \sigma_n$ are n neurons, with $\sigma_i = (v_i, R_i, (r_i, c_i))$, $i = 1, 2, \ldots, n$, wherein
   a. $v_i \in \mathbb{R}$ is the state of σ_i, wherein $\mathbb{R}$ is the real number set;
   b. R_i is the set of nonlinear spiking rules, $\alpha^{g(v)} \rightarrow \alpha^{f(v)}$, in which functions $g(\cdot)$ and $f(\cdot)$ are linear or nonlinear;
   c. $r_i = \rho_{i1}(x_i, v_i)$ and $c_i = \rho_{i2}(x_i, v_i)$ are nonlinear gates, wherein $x_i \in \mathbb{R}$ is the external input;
3. $syn \subseteq \{1, 2, \ldots, n\} \times \{1, 2, \ldots, n\}$, for all $(i, j) \in syn$, $i \neq j$, $i, j = 1, 2, \ldots, n$ (synapses).

From a topological perspective, the NGSNP-II architecture can be modeled as a digraph comprising gated neuronal units as vertices, with directed edges representing inter-neuronal synaptic connections. Illustrated in Figure 3.6a is a nonlinear spiking unit σ_i within the NGSNP-II framework. Every neural unit contains a data component representing its internal state. The value $v_i(\tau)$ indicates the state of neuron σ_i at moment τ. The spiking dynamics follow a nonlinear rule expressed as $\alpha^{g(v)} \rightarrow \alpha^{f(v)}$, with $f(\cdot)$ and $g(\cdot)$ representing arbitrary linear or nonlinear functional mappings. When the functions $f(\cdot)$ and $g(\cdot)$ exhibit linearity, the NGSNP-II model operates identically to a standard SNP system. This equivalence positions

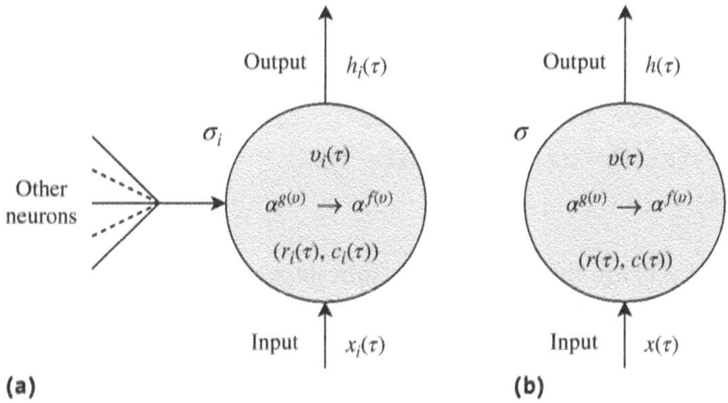

Figure 3.6 (a) In a NGSNP-II system, neuron σ_i and its presynaptic neurons are shown; (b) only one neuron σ is designed in a specialized NGSNP-II system. Adapted from [5].

NGSNP-II as a generalized framework incorporating nonlinear dynamics into traditional SNP architectures. Usually, $g(\cdot)$ and $f(\cdot)$ are different functions, and can also be the same. If $f(\cdot) \equiv 0$, the rule is known as the forgetting rule.

The semantics of rule $\alpha^{g(v)} \rightarrow \alpha^{f(v)}$ is explained as below. If σ_i fires at timestep τ, then $g(v)$ spikes are removed, and $f(v)$ spikes are produced and then transmitted to the postsynaptic neurons. Thus, $f(\cdot)$ and $g(\cdot)$ are known as spike generation and spike consumption functions, respectively. According to the spiking semantics, the state equation can be given by

$$v_i(\tau) = r_i(\tau) \cdot v_i(\tau - 1) - c_i(\tau) \cdot g(v) + m \tag{3.23}$$

where $v = x_i(\tau) + v_i(\tau - 1)$; $x_i(\tau)$ is an external input at time τ; $v_i(\tau - 1)$ is the state of gated neuron σ_i at time $\tau - 1$, and m is the value of spikes from its presynaptic neurons; $r_i(\tau)$ and $c_i(\tau)$ are nonlinear gate functions. Furthermore, the mathematical model connecting inputs to outputs is given by

$$h(\tau) = f(u) = f(v_i(\tau - 1) + x_i(\tau)) \tag{3.24}$$

wherein $x_i(\tau)$ and $v_i(\tau - 1)$ have the same meaning as before.

3.4.2 Gated Spiking Neural P

In the context of time series prediction, the model employs a single gating unit represented by σ, while treating the consumption and generation functions as identical (i.e., $g(\cdot) \equiv f(\cdot)$).

Figure 3.6b shows the neuron σ in the specialized NGSNP-II system. Since only one gated neuron is considered in Eq. (3.23), $m = 0$. Consequently, the state and input–output equations for σ are reformulated by

$$\begin{cases} v(\tau) = r(\tau) \cdot v(\tau - 1) - c(\tau) \cdot f(x(\tau) + v(\tau - 1)) \\ h(\tau) = f(x(\tau) + v(\tau - 1)) \end{cases} \tag{3.25}$$

wherein $v(\tau - 1)$ and $v(\tau)$ are the states of σ at timesteps $\tau - 1$ and τ; $h(\tau)$ is its output at timestep τ, and $x(\tau)$ is an external input at timestep τ; functions $r(\tau)$ and $c(\tau)$ are nonlinear.

The GSNP framework [5], corresponding to Eq. (3.25), constitutes a particular case of the NGSNP-II network where the system reduces to merely one gating unit σ, visualized in Figure 3.7.

The gated unit σ exhibits properties resembling recurrent neural architectures. Temporal unfolding reveals that this σ-based gated structure (i.e., the GSNP framework) demonstrates recurrence characteristics, as illustrated in Figure 3.8. At every moment, the neuron has an external input, and it can calculate an output according to the state at the last time and the current input. And two gates are used to update the state of neuron. Therefore, GSNP model is suitable for time series forecasting problems.

For comprehensive modeling purposes, the GSNP system's scalar elements can be elevated to vector status across all dimensions, with $\mathbf{x}(\tau)$ characterizing inputs, $\mathbf{v}(\tau)$ describing states, and $\mathbf{h}(\tau)$ representing outputs in this generalized formulation. Consequently, the state equation and output relation can be reformulated as

$$\begin{cases} \mathbf{v}(\tau) = r(\tau) \odot \mathbf{v}(\tau - 1) - c(\tau) \odot f(W_x \cdot \mathbf{x}(\tau) + W_v \cdot \mathbf{v}(\tau - 1) + \mathbf{b}) \\ \mathbf{h}(\tau) = W_h \cdot f(W'_x \cdot \mathbf{x}(\tau) + W'_v \cdot \mathbf{v}(\tau - 1) + \mathbf{b}') \end{cases} \tag{3.26}$$

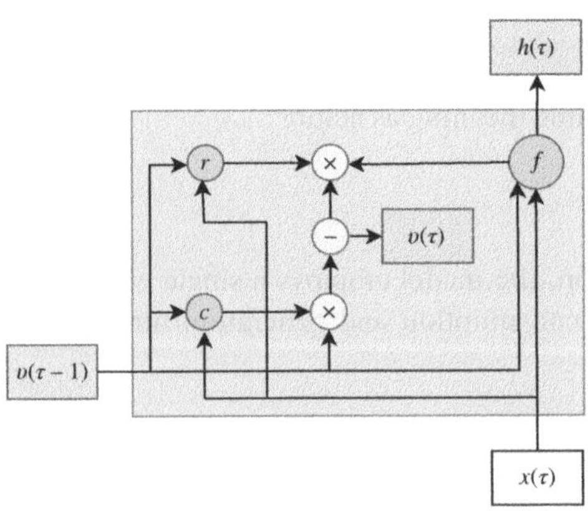

Figure 3.7 Functional structure of GSNP model [5]/IEEE.

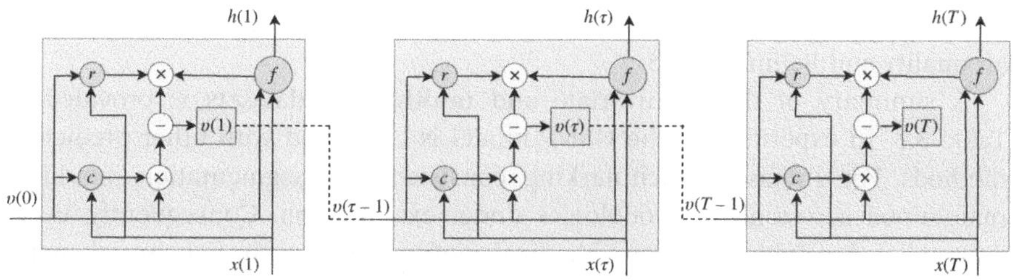

Figure 3.8 GSNP model for predicting time series. Adapted from [5].

wherein W_x, W_v, W_h, W_x', and W_v' are the matrices; $\mathbf{b}$ and $\mathbf{b}'$ are the biases; $\odot$ denotes the inner product.

Equation (3.26) introduces two gating mechanisms: consumption gate $c(\tau)$ and reset gate $r(\tau)$, which regulate the recurrent architecture depicted in Figure 3.8. These gates are given by

1. "Reset gate": It regulates the degree to which prior information is overwritten, based on present inputs, historical data, and systematic offsets.

$$r(\tau) = \rho(V_r \cdot \mathbf{v}(\tau - 1) + W_r \cdot \mathbf{x}(\tau) + \mathbf{b}_r) \tag{3.27}$$

2. "Consumption gate": It calculates the proportion of the prior state utilized according to the present input, historical state, and bias.

$$c(\tau) = \rho(V_c \cdot \mathbf{v}(\tau - 1) + W_c \cdot \mathbf{x}(\tau) + \mathbf{b}_c) \tag{3.28}$$

wherein $\mathbf{b}_r$ and $\mathbf{b}_c$ are biases; W_r, V_r, W_c, and V_c are parameter matrices; function $\rho(\cdot)$ is nonlinear.

The GSNP model is expressed by Eqs. (3.26)–(3.28), where W_x, W_v, W_h, W_x', W_v', W_r, V_r, W_c, V_c, $\mathbf{b}_c$, $\mathbf{b}_r$, $\mathbf{b}'$, and $\mathbf{b}$ are the parameters to be trained. Thus, the GSNP model is a recurrent-like model, shown in Figure 3.8. Consequently, the standard BPTT approach remains applicable for training GSNP architectures.

3.4.3 Model Evaluation

3.4.3.1 Datasets

The GSNP model's performance is assessed using a collection of ten real-world datasets, comprising eight single-variable and two multivariable sequences.

The single-variable sequences are CO_2 (CO_2 [ppm] at Mauna Loa), Lake (monthly Lake Erie levels), Closing (monthly closings of the Dow-Jones industrial index), Radio (monthly critical radio frequencies), Sunspot (Sunspot time series), Milk (monthly milk production per cow), SP500 (S&P 500 stock index), and MG (Mackey–Glass time series). Two multivariable sequences include Italian air quality and Beijing PM2.5.

A summary of these univariate and multivariate datasets is provided in Table 3.3. In experiments, the GSNP model is compared with other prediction methods. For unbiased benchmarking, consistent data segmentation should be maintained across all methodologies under examination. Consequently, every dataset listed in Table 3.3 follows identical configurations as the benchmark methods. Additionally, all time-series features are normalized to the interval $[-1, 1]$.

Two distinct experimental configurations are established, including single-variable and multivariable temporal sequences. These testing environments facilitate comprehensive comparisons between the proposed GSNP architecture and existing predictive models ranging from fundamental references to advanced solutions.

3.4.3.2 Performance Metrics

Two measurement indicators are used to analyze the performance of these forecasting methods: RMSE and MAE (mean absolute error), and expressed by

$$\mathrm{RMSE} = \sqrt{\frac{1}{MT}\sum_{i=1}^{M}\sum_{\tau=1}^{T}(x_i(\tau) - \tilde{x}_i(\tau))^2} \tag{3.29}$$

$$\mathrm{MAE} = \frac{1}{MT}\sum_{i=1}^{M}\sum_{\tau=1}^{T}|x_i(\tau) - \tilde{x}_i(\tau)| \tag{3.30}$$

Table 3.3 Statistical information of datasets. Adapted from [5].

Name	Variables number	Size	Test size	Training size
SP500	1	251	60	191
MG	1	1,000	60	940
Sunspot	1	288	60	228
Milk	1	168	60	108
Lake	1	600	60	540
Closing	1	291	60	231
Radio	1	240	60	180
CO_2	1	192	60	132
Beijing PM2.5	9	43,824	8,765	35,059
Italian air quality	15	9,358	1,872	7,486

wherein the parameter M denotes variable count (with $M = 1$ for univariate cases), T represents temporal duration, while x_i and $\widetilde{x}_i$ correspond to observed and forecasted sequences respectively.

3.4.3.3 Univariate Datasets

For evaluating GSNP's predictive capability, six state-of-the-art forecasting approaches serve as benchmark models: WAR [7], MAR [8], ANN [9], TCN [10], CNN-FCM [14], and ANFIS [11]. ANFIS represents a fuzzy reasoning system built upon adaptive neural networks. The WAR model employs wavelet transform in its autoregressive framework, whereas MAR adopts a multiresolution approach for autoregressive modeling. ANN utilizes neural networks for predictive tasks, while CNN-FCM integrates convolutional neural networks with fuzzy cognitive mapping. In contrast, TCN employs a hybrid architecture of convolutional and recurrent networks.

Table 3.4 provides a prediction comparison of these models for eight univariate time series in terms of RMSE metric. Table 3.4 demonstrates distinct predictive capabilities across models: (i) The GSNP model outperforms others for SP500, Closing, and CO_2 datasets; (ii) ANFIS shows superior accuracy specifically for Milk series; (iii) MAR exhibits optimal performance on Lake data; (iv) CNN-FCM dominates in Sunspot and Radio predictions; (v) TCN provides the most accurate forecasts for MG series.

3.4.3.4 Multivariate Datasets

Two standard multivariate time series datasets are used: (i) Italian air quality and (ii) Beijing PM2.5. Table 3.3 provides an overview of these datasets, while comprehensive details are available in Ref. [19]. The experimental setup allocates the initial 80% of data for training purposes, reserving the remaining 20% for testing.

Table 3.4 The RMSE values of the different methods on univariate datasets. Adapted from [5].

Datasets	GSNP	ANFIS	WAR	MAR	ANN	CNN-FCM	TCN
MG	0.0013	0.0010	0.0350	0.0020	0.0050	0.0013	0.0008
SP500	11.2057	14.9350	17.7970	16.0410	17.6960	20.8158	51.2674
SunSpot	22.8846	22.7530	35.2620	19.1860	19.9010	17.9488	22.4489
Milk	38.5390	9.5780	57.7170	37.8380	27.1130	30.4738	33.8579
Lake	0.4605	0.4580	0.6380	0.3900	0.4020	0.3913	0.4093
Closing	24.3763	27.5620	29.8220	26.7330	28.5320	25.1895	25.2143
Radio	0.6882	0.6510	0.9020	0.6620	0.6520	0.5665	0.6020
CO_2	0.4248	0.9010	1.3500	1.6950	1.6950	0.7305	3.1202

Two multivariate temporal sequences are employed to benchmark the GSNP approach alongside a set of 10 comparative models. The comparative models include VARMA [20], ARIMA [20], RNN [21], SVR-LIN [22], SVR-RBF [22], CNN [21], SEQ2SEQ [23], SEQ2SEQ-BI [23], SEQ2SEQ-ATT [23], and MTSMFF [19]. Ref. [19] contains complete implementation details and performance metrics of all benchmarked approaches.

Table 3.5 displays the RMSE and MAE metric evaluations contrasting the GSNP framework against ten reference models across two multivariate temporal datasets. Moreover, Table 3.6 provides the comparison of the GSNP model with 10 baseline models for Beijing PM2.5 multistep forecasting in terms of RMSE metric. The purpose of the second comparison is to evaluate the ability of GSNP and 10 baseline methods for multistep forecasting task. The results of the 10 baseline methods are retrieved from Ref. [19]. Table 3.5 demonstrates that the GSNP model has superior forecasting accuracy compared to alternative models when applied to Beijing's PM2.5 multivariate time series data. Nevertheless, its performance degrades significantly for Italian air quality data, as evidenced by elevated RMSE and MAE metrics, suggesting inadequate adaptation to this dataset's variations. Consequently, further enhancements are required to optimize the model's predictive capabilities.

Table 3.5 The RMSE and MAE values of the different methods on two multivariate datasets [5]/IEEE.

	Beijing PM2.5		Italian air quality	
Methods	**RMSE**	**MAE**	**RMSE**	**MAE**
GSNP	11.20	3.34	99.34	44.72
MTSMFF	39.83	25.45	53.28	40.25
ARIMA	53.41	41.06	59.07	50.66
VARMA	51.78	38.95	58.44	47.32
SVR-LIN	88.35	71.48	72.39	65.48
SVR-RBF	57.64	43.47	59.02	48.49
RNN	44.42	27.99	53.35	40.16
CNN	42.98	26.89	54.13	41.07
SEQ2SEQ	42.53	26.42	54.89	41.25
SEQ2SEQ-BI	42.29	26.31	54.45	41.21
SEQ2SEQ-ATT	42.15	26.29	54.22	41.18

Table 3.6 The RMSE values of the different methods on Beijing PM2.5 multistep forecasting tasks [5]/IEEE.

Method	$t+1$	$t+2$	$t+3$	$t+4$	$t+5$	$t+6$
GSNP	10.70	10.70	10.73	13.69	10.70	10.71
SVR-RBF	45.02	49.97	55.29	61.66	64.52	69.41
SVR-LIN	38.73	47.39	66.30	163.5	120.34	93.89
VARMA	24.10	40.05	55.27	60.36	62.46	68.47
ARIMA	24.21	41.17	56.43	62.96	65.28	70.38
MTSMFF	23.52	33.06	41.09	47.12	48.05	46.15
RNN	24.93	35.23	43.27	50.11	54.49	58.48
CNN	24.57	34.26	41.98	48.02	52.68	56.37
SEQ2SEQ	24.42	33.36	41.83	48.07	52.17	55.34
SEQ2SEQ-BI	23.86	33.42	42.05	48.13	52.11	54.20
SEQ2SEQ-ATT	23.94	33.87	42.33	48.09	51.25	53.41

3.5 NSNP-AU Model

3.5.1 Type-III Nonlinear Gated Spiking Neural P Systems

In this section, the third gated variant of nonlinear spiking neural P systems is discussed, termed type-III nonlinear gated spiking neural P systems (NGSNP-III systems). The NGSNP-III systems are also known as nonlinear spiking neural P systems with autapses (NSNP-AU systems), because they can be viewed as nonlinear variant of spiking neural P systems with autapses.

Definition 3.3 A NGSNP-III system (or NSNP-AU system) is defined by

$$\Pi = (O, \sigma_1, \sigma_2, \ldots, \sigma_n, syn)$$

where

1. $O = \{\alpha\}$ is the singleton alphabet, where α is known as the spike;
2. $\sigma_1, \sigma_2, \ldots, \sigma_n$ are n neurons with autapses, of the form $\sigma_i = (v_i, R_i, (r_i, c_i, o_i))$, $i = 1, 2, \ldots, n$, wherein
   a. $v_i \in \mathbb{R}$ denotes the state of σ_i, in which $\mathbb{R}$ is the real number set;
   b. R_i is the set of spiking rules, $\alpha^{g(v)} \rightarrow \alpha^{f(v)}$, where functions $g(\cdot)$ and $f(\cdot)$ are linear or nonlinear;

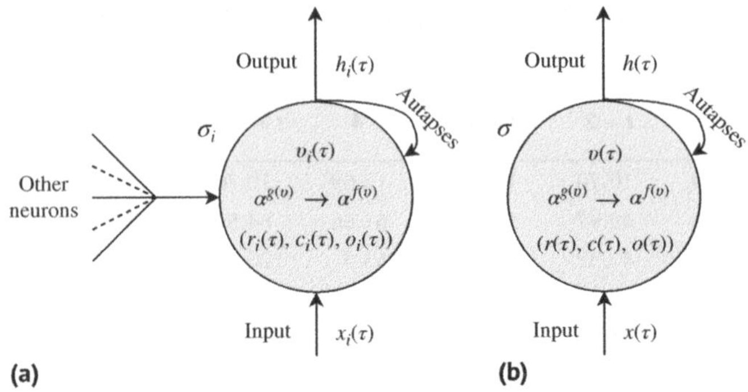

Figure 3.9 (a) In a NGSNP-III system (or NSNP-AU system), neuron σ_i and its presynaptic neurons are shown; (b) Only one neuron σ is designed in a specialized NGSNP-III system [6]/IEEE.

c. $r_i = \rho_{i1}(x_i, v_i, h_i)$, $c_i = \rho_{i2}(x_i, v_i, h_i)$ and $o_i = \rho_{i3}(x_i, v_i, h_i)$ are nonlinear gates, in which $x_i \in \mathbb{R}$ is the external input, and $h_i \in \mathbb{R}$ is the output;

3. $syn \subseteq \{1, 2, \ldots, n\} \times \{1, 2, \ldots, n\}$ (synapses).

A NGSNP-III system (or NSNP-AU system) is expressed by a directed graph of neurons with autapses, in which the arcs indicate the synapses between these neurons. Each neuron has an internal state unit, and the state at time τ is denoted by $v_i(\tau)$. Moreover, the nonlinear spiking rule has the form $a^{g(v)} \to a^{f(v)}$. The autaptic neuron σ_i is depicted in Figure 3.9a. Under the condition where $f(\cdot)$ and $g(\cdot)$ maintain linearity, the spike generation follows traditional patterns. This establishes NGSNP-III/NSNP-AU systems as nonlinear generalizations of their SNP-AU counterparts. A key differentiation between NGSNP-III and SNP-AU systems lies in their gating structures, with the former employing three distinct nonlinear gate operations: $r_i = \rho_{i1}(x_i, v_i, h_i)$, $c_i = \rho_{i2}(x_i, v_i, h_i)$ and $o_i = \rho_{i3}(x_i, v_i, h_i)$. These gating mechanisms serve as regulatory tools to achieve a recurrent-like architecture, specifically governing the emission and absorption of neural spikes.

The nonlinear spiking rule includes two mechanisms: nonlinear spike consumption and nonlinear spike generation. When neuron σ_i activates, it utilizes spikes valued at $g(v)$ while simultaneously producing new spikes valued at $f(v)$, which are subsequently transmitted to downstream neurons. Consequently, $g(\cdot)$ and $f(\cdot)$ are designated as the spike consumption and production functions, respectively. Following spiking neural dynamics, state transition of σ_i is mathematically formulated as

$$v_i(\tau) = r_i(\tau) \cdot v_i(\tau - 1) - c_i(\tau) \cdot g(v_i(\tau - 1) + x_i(\tau) + m) \tag{3.31}$$

where $r_i(\tau) = \rho_{i1}(v_i(\tau - 1) + h_i(\tau - 1) + x_i(\tau))$, and $c_i(\tau) = \rho_{i2}(h_i(\tau - 1) + x_i(\tau) + v_i(\tau - 1))$; $v_i(\tau - 1)$ and $v_i(\tau)$ are the states of σ_i at timesteps $\tau - 1$ and τ; $x_i(\tau)$ is an external input at timestep τ, $h_i(\tau-1)$ is the output of σ_i at timestep $\tau-1$, and m is the value of spikes received from its presynaptic neurons. Additionally, the output is computed by

$$h_i(\tau) = o_i(\tau) \cdot f(v_i(\tau-1) + x_i(\tau)) \tag{3.32}$$

wherein $o_i(\tau) = \rho_{i3}(x_i(\tau) + h_i(\tau-1) + v_i(\tau-1))$; $h_i(\tau)$ denotes the output, and $o_i(\tau)$ is nonlinear. The gate operations $r_i(\tau)$, $c_i(\tau)$, and $o_i(\tau)$ are influenced not only by $x_i(\tau)$ and $v_i(\tau-1)$ but also by the prior hidden state $h_i(\tau-1)$, as illustrated in Figure 3.9a through autaptic connections.

The operational procedure of an NGSNP-III/NSNP-AU system initiates from predefined starting conditions and proceeds iteratively until termination criteria are satisfied. Throughout this computational cycle, individual neuronal states undergo transformation following Eq. (3.31), while their corresponding outputs are determined by Eq. (3.32).

3.5.2 NSNP-AU Model

For sequence data analysis, the model adopts two simplifications: (i) a single neural unit σ is modeled; (ii) identical functional forms are assumed for both spike generation and consumption $g(\cdot) \equiv f(\cdot)$. The neuron σ is shown in Figure 3.9b. In the scenario where merely one neuron σ exists, m is set to 0 in Eq. (3.31). As a result, the dynamical representation of neuron σ is redefined by revising its state/output equations:

$$\begin{cases} v(\tau) = r(\tau) \cdot v(\tau-1) - c(\tau) \cdot g(x(\tau) + v(\tau-1)) \\ h(\tau) = o(\tau) \cdot f(x(\tau) + v(\tau-1)) \end{cases} \tag{3.33}$$

wherein $x(\tau)$ denotes the external input at moment τ; $h(\tau-1)$ presents the output of neuron σ at moment $\tau-1$; $v(\tau)$ and $v(\tau-1)$, respectively, are the states of σ at moments τ and $\tau-1$; $h(\tau)$ is the output of σ at moment τ; $o(\tau)$, $c(\tau)$ and $r(\tau)$ are three nonlinear gate functions: $r(\tau) = \rho_1(v(\tau-1) + h(\tau-1) + x(\tau))$, $c(\tau) = \rho_2(v(\tau-1) + h(\tau-1) + x(\tau))$, and $o(\tau) = \rho_3(v(\tau-1) + h(\tau-1) + x(\tau))$.

The neuron σ exhibits recurrent characteristics and is designated as a NSNP-AU neuron. The computational workflow described in Eq. (3.33) is illustrated through the NSNP-AU neuron's functional architecture in Figure 3.10.

Figure 3.11 presents the NSNP-AU model [6], which is formed by unfolding the NSNP-AU neuron temporally to create a recurrent-like computational framework. The NSNP-AU neuron receives an external input at every time step, generating an output determined by both the present input and its prior state, regulated by the gating function $o(\tau)$. And the state of the neuron is updated based on the current input, and the state and output at the last time are determined by two gate

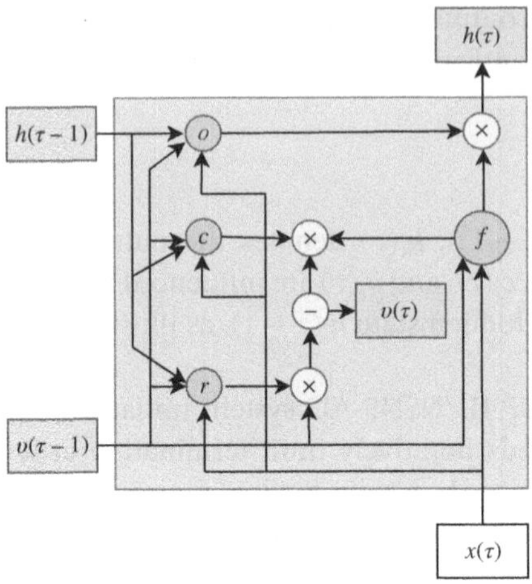

Figure 3.10 Functional structure of NSNP-AU model [6]/IEEE.

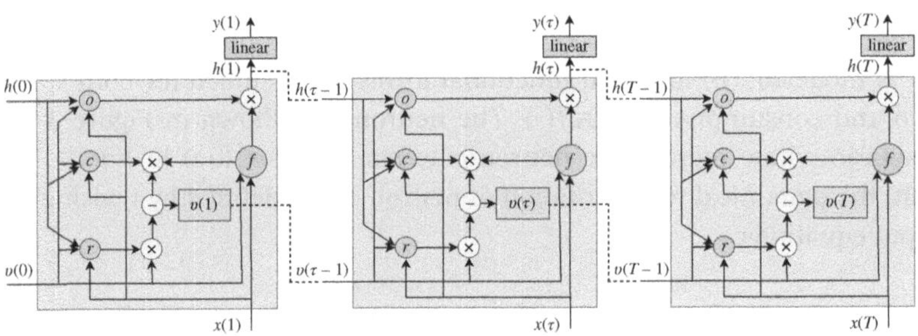

Figure 3.11 NSNP-AU model [6]/IEEE.

functions, $r(\tau)$ and $c(\tau)$. Therefore, the NSNP-AU model seems suitable for problems related to sequence data.

$h(\tau)$ is the output of the NSNP-AU neuron. The NSNP-AU model's architecture is enhanced by adding a linear projection layer during training, which modifies the output from $h(\tau)$ to $y(\tau)$.

The NSNP-AU architecture is expanded into vector space for comprehensive modeling, with multidimensional input $\mathbf{x}(\tau)$, output $\mathbf{y}(\tau)$, and state variables $\mathbf{h}(\tau)$ and $\mathbf{v}(\tau)$. As a result, the formulation of both output and state equations is modified by

$$\begin{cases} \mathbf{v}(\tau) = r(\tau) \odot \mathbf{v}(\tau - 1) - c(\tau) \odot f(W_x \cdot \mathbf{x}(\tau) + W_v \cdot \mathbf{v}(\tau - 1) + \mathbf{b}) \\ \mathbf{y}(\tau) = W_y \cdot o(\tau) \odot f(W'_x \cdot \mathbf{x}(v) + W'_v \cdot \mathbf{v}(\tau - 1) + \mathbf{b}') \end{cases} \tag{3.34}$$

where $\mathbf{b}'$ and $\mathbf{b}$ are the biases; W_x, W_v, W_y, W'_x, and W'_v are the matrices; "$\odot$" denotes the inner product.

Equation (3.34) introduces three gating mechanisms: generation gate $o(\tau)$, consumption gate $c(\tau)$, and reset gate $r(\tau)$. These components regulate the recurrent architecture depicted in Figure 3.10. These gates are given by

$$\begin{cases} r(\tau) = \rho_1(\tau)(W_r \cdot \mathbf{x}(\tau) + V_r \cdot \mathbf{v}(\tau-1) + H_r \cdot \mathbf{h}(\tau-1) + \mathbf{b}_r) \\ c(\tau) = \rho_2(\tau)(W_c \cdot \mathbf{x}(\tau) + V_c \cdot \mathbf{v}(\tau-1) + H_c \cdot \mathbf{h}(\tau-1) + \mathbf{b}_c) \\ o(\tau) = \rho_3(\tau)(W_o \cdot \mathbf{x}(\tau) + V_o \cdot \mathbf{v}(\tau-1) + H_o \cdot \mathbf{h}(\tau-1) + \mathbf{b}_o) \end{cases} \tag{3.35}$$

wherein $\mathbf{b}_c$, $\mathbf{b}_r$, and $\mathbf{b}_o$ are bias vectors, and V_c, W_c, V_r, W_r, H_r, H_c, W_o, V_o, and H_o are the matrices.

The NSNP-AU model is described by Eqs. (3.34) and (3.35), where W_x, W_v, W_y, W'_x, W'_v, W_r, V_r, H_r, W_c, V_c, H_c, W_o, V_o, and H_o, $\mathbf{b}$, $\mathbf{b}'$, $\mathbf{b}_r$, $\mathbf{b}_c$, and $\mathbf{b}_o$ are the parameters to be trained. The NSNP-AU model is a recurrent-like model and can be regarded as a RNN variant. Therefore, the usual BPTT algorithm can be used to train the NSNP-AU model.

3.5.3 Model Evaluation

The experimental validation of the NSNP-AU model's forecasting capability employs four standard chaotic time series datasets. The four datasets include Gas Furnace, Sunspot, Lorenz, and Mackey–Glass time series.

The NSNP-AU model is evaluated against multiple advanced and baseline prediction models using three performance metrics: symmetric mean absolute percentage error (SMAPE), normalized root mean square error (NRMSE), and RMSE. The RMSE measure has been defined in Eq. (3.22). The SMAPE and NRMSE measures are given by

$$\text{NRMSE} = \sqrt{\frac{1}{T\sigma^2} \sum_{\tau=1}^{T} (x_i(t) - \tilde{x}_i(\tau))^2} \tag{3.36}$$

$$\text{SMAPE} = \frac{2}{T} \sum_{\tau=1}^{T} \frac{|x_i(\tau) - \tilde{x}_i(\tau)|}{|x_i(\tau)| + |\tilde{x}_i(\tau)|} \tag{3.37}$$

where T is the number of data samples; $x_i(\tau)$ and $\tilde{x}_i(\tau)$ are the original and the estimated samples; and σ^2 denotes the variance of samples.

3.5.3.1 Lorenz Dataset

A benchmark task is to predict Lorenz chaotic time series. The Lorenz system is given by

$$\begin{cases} \frac{dx}{dt} = a(-x+y) \\ \frac{dy}{dt} = bx - y - xz \\ \frac{dz}{dt} = xy - cz \end{cases} \tag{3.38}$$

To maintain the chaotic behavior, the chosen parameters for the Lorenz system are $c = 3/8$, $b = 28$, and $a = 10$ (see Ref. [24]). Based on the step size and initial condition, Lorenz time series is generated by Runge–Kutta method. However, the different step sizes and initial conditions can generate the different Lorenz time series. Thus, three cases are studied as follows.

(1) Case I:
The step size is 0.01, and initial condition is $(x(0), y(0), z(0)) = (1, 1, 0)$. A total of 5,000 data points are created based on the step size and initial condition, with 3,000 allocated for model training and 2,000 reserved for testing purposes. Each data point is normalized to fall within the $[-1, 1]$ range.

In Table 3.7, the performance of NSNP-AU is evaluated against multiple benchmark approaches using Lorenz chaotic system data (Case I). The benchmark approaches are BPSO-ESN [25], Lasso-ESN [26], RR-ESN [27], DESN [28], SCR [29], OESN [30], and ALM-ESN [31]. Performance metrics for comparison approaches are obtained from Ref. [31]. According to Table 3.7, both ALM-ESN and NSNP-AU outperform the remaining six benchmark approaches, with NSNP-AU demonstrating marginally superior performance to ALM-ESN. These findings confirm the adequacy of the NSNP-AU framework for Lorenz chaotic dataset.

(2) Case II:
The step size is 0.02 and initial condition is $(x(0), y(0), z(0)) = (12, 2, 9)$. In Case II, Lorenz dataset has 1,500 points, using the first 1,000 points as the training set and the remaining 500 samples as testing. All points are normalized to $[-1, 1]$.

In Table 3.8, NSNP-AU is evaluated against multiple benchmark approaches using Lorenz chaotic system data (Case II). The benchmark approaches are ESGP, ESN with Bayesian regression, RESN, SVESM with a quadratic loss function, SVESM with the Huber loss function. Performance metrics for comparison approaches are obtained from Ref. [32]. The evaluation demonstrates NSNP-AU's superior performance relative to comparison approaches.

Table 3.7 The results of the different models for Case I [6].

Models	NRMSE
NSNP-AU	1.11e-5 (±1.41e-11)
ALM-ESN	2.21e-5 (±3.21e-6)
OESN	9.19e-4 (±9.21e-5)
SCR	9.92e-4 (±8.31e-5)
DESN	7.72e-4 (±7.45e-5)
RR-ESN	8.38e-4 (±8.13e-5)
Lasso-ESN	5.19e-4 (±7.16e-5)
BPSO-ESN	5.23e-4 (±6.31e-5)

Table 3.8 The results of the different models for Case II [6]/IEEE.

Models	RMSE
ESGP	0.4386
ESN with Bayesian regression	0.4192
RESN	0.2238
SVESM with quadratic loss function	0.3714
SVESM with Huber loss function	0.2644
NSNP-AU	0.1869

(3) Case III:

The Lorenz system generates multivariate chaotic time series data. Beyond analyzing $x(t)$, researchers like those in Ref. [33] have demonstrated forecasting capabilities for the $y(t)$ component. Case III examines the forecasting of $y(t)$ in the Lorenz chaotic system, with a step size of 0.01 and initial values set to $(x(0), y(0), z(0)) = (12, 2, 9)$. For the third case, the Lorenz dataset comprises 2,501 data points, with 2,001 allocated to training and the subsequent 500 reserved for validation. Each data point is normalized to the range of $[-1, 1]$.

Table 3.9 gives the comparisons of NSNP-AU with other forecasting models on the Lorenz dataset of Case III. These models are ANS-QKRLS [33], EX-KRLS [34], SW-KRLS [35], ALD-KRLS [36], FB-KRLS [37], KRLS-T [38], and QKRLS [39]. Performance metrics for comparison approaches are obtained from Ref. [33]. As shown in Table 3.9, both ANS-QKRLS and NSNP-AU outperform the remaining six algorithms in the comparison. These findings confirm that NSNP-AU possesses adequate predictive capability for Lorenz chaotic time series forecasting.

Table 3.9 The results of the different models for Case III [6]/IEEE.

Models	RMSE
QKRLS	0.1767 (±0.0208)
KRLS-T	0.1133 (±0.0467)
FB-KRLS	0.1981 (±0.1096)
EX-KRLS	0.9237 (±0.1425)
SW-KRLS	1.1487 (±0.3310)
ALD-KRLS	0.0876 (±0.0282)
ANS-QKRLS	0.0441 (±0.0208)
NSNP-AU	0.0660 (±1.9039e-4)

3.5.3.2 Mackey–Glass Dataset

The NSNP-AU model's performance is assessed using Mackey–Glass chaotic dataset, which is derived from a specific delay differential equation [40]:

$$\frac{dx(t)}{dt} = -0.1x(t) + \frac{0.2x(x-\tau)}{1+x^{10}(t-\tau)} \tag{3.39}$$

wherein $17 \leq \tau$. Equation (3.39) can show chaotic behaviors.

The MG dataset has 1,000 samples, using the first 500 samples for training and the remaining 500 samples for testing. All samples are produced based on $\tau = 17$ and normalized to $[-1, 1]$.

The performance of NSNP-AU against multiple benchmark prediction approaches is evaluated in Table 3.10 using the Mackey–Glass dataset. The benchmark prediction approaches are MSBFNN [41], RBF-AFS [42], OLS [42], ANFIS [42], ILA [42], D-FNN [43], SEIT2FNN [43], G-FNN [43], NEFPROX [43], and ProFNN [44]. Performance metrics for comparison approaches are obtained from Ref. [44]. As evidenced in Table 3.10, the NSNP-AU approach outperforms all alternative forecasting techniques. The experimental findings confirm its superior capability in modeling the Mackey–Glass chaotic system.

3.5.3.3 Sunspot Dataset

The sunspot activity data spanning from January 1749 to June 2013 are utilized, containing 3,174 annual observations. For experimental purposes, the dataset is partitioned into training (initial 2,000 data points) and testing (remaining 1,174 points) subsets. All values are scaled to the range of $[-1, 1]$ for normalization.

Table 3.11 presents a comparative analysis between NSNP-AU and seven alternative methods utilizing echo state networks for Sunspot prediction. The alternative methods are BPSO-ESN [25], Lasso-ESN [26], RR-ESN [27], DESN [28],

Table 3.10 The results of the distinct models on Mackey–Glass dataset [6]/IEEE.

Models	RMSE
ProFNN	0.0027
NEFPROX	0.0533
G-FNN	0.0056
SEIT2FNN	0.0034
D-FNN	0.0080
ILA	0.0066
ANFIS	0.0030
OLS	0.0089
RBF-AFS	0.0128
MSBFNN	0.0024
NSNP-AU	0.0023

Table 3.11 The results of the different methods on Sunspot dataset [6]/IEEE.

Models	NRMSE
BPSO-ESN	0.1923 (±0.0276)
Lasso-ESN	0.1902 (±0.0232)
RR-ESN	0.2135 (±0.0289)
DESN	0.2310 (±0.0325)
SCR	0.2604 (±0.0438)
OESN	0.2468 (±0.0424)
ALM-ESN	0.1611 (±0.0133)
NSNP-AU	0.1808 (±3.3154e-7)

SCR [29], OESN [30], and ALM-ESN [31]. Performance metrics for comparison approaches are obtained from Ref. [31]. As shown in Table 3.11, (i) ALM-ESN demonstrates superior predictive capability, while NSNP-AU ranks second with marginally inferior performance compared to the top model; (ii) NSNP-AU model achieves the smallest variance. The comparisons demonstrate that NSNP-AU model is sufficient for predicting Sunspot data.

3.5.3.4 Gas Furnace Dataset

The Gas Furnace benchmark dataset represents a commonly employed multivariate temporal sequence in machine learning applications. This collection features

Table 3.12 The results of the different methods on Gas Furnace dataset [6]/IEEE.

Models	RMSE
Similarity degree of GRA	0.9072 (±6.2382e-4)
IGRA based on area	0.4700 (±5.1699e-4)
Deng's GRA	0.5200 (±5.1699e-4)
Han's IGRA	0.4423 (±5.0497e-4)
NSNP-AU	0.3648 (±1.3906e-4)

gas flow rate $u(t)$ as the input parameter while measuring CO_2 concentration percentage $y(t)$ as the output. Comprising 290 data points, the collection is partitioned into 200 samples for model training and 90 for validation purposes. All values undergo normalization to the interval between −1 and 1.

The comparative evaluation of NSNP-AU against multiple GRA-based approaches for Gas Furnace time series forecasting is presented in Table 3.12.

Table 3.12 compares NSNP-AU with several methods based on grey relational analysis (GRA) for Gas Furnace data forecasting. The comparison methods are IGRA based on area [45], the similarity degree of GRA [46], Deng's GRA [47], and Han's IGRA [48]. Performance metrics for comparison approaches are obtained from Ref. [48]. Experimental evaluations demonstrate that NSNP-AU attains the minimal average RMSE with the least fluctuation among all benchmarked approaches, establishing its superior predictive capability for the Gas Furnace chaotic time series.

References

1 S. Hochreiter and J. Schmidhuber. Long short-term memory. *Neural Computation*, 9(8): 1735–1780, 1997.

2 F.A. Gers, J. Schmidhuber, and F. Cummins. Learning to forget: continual prediction with LSTM. *1999 Ninth International Conference on Artificial Neural Networks (ICANN 99)*, 1999.

3 R. Dey and F.M. Salemt. Gate-variant of gated recurrent unit (GRU) neural network. *2017 IEEE 60th International Midwest Symposium on Circuits and Systems*, pages 1597–1600, 2017.

4 Q. Liu, L. Long, Q. Yang, H. Peng, J. Wang, and X. Luo. LSTM-SNP: A long short-term memory model inspired from spiking neural P systems. *Knowledge-Based Systems*, 235: 107656, 2022.

5 Q. Liu, L. Long, H. Peng, J. Wang, Q. Yang, X. Song, A. Riscos-Núñez, and M.J. Pérenz-Jiménez. Gated spiking neural P systems for time series forecasting. *IEEE Transactions on Neural Networks and Learning Systems*, 34(9): 6227–6236, 2023.
6 Q. Liu, H. Peng, L. Long, J. Wang, Q. Yang, M.J. Pérez-Jiménez, and D. Orellana-Martín. Nonlinear spiking neural systems with autapses for predicting chaotic time series. *IEEE Transactions on Cybernetics*, 54(3): 1841–1853, 2024.
7 G. Zheng, J. Starck, J. Campbell, and F. Murtagh. Multiscale transforms for filtering financial data streams. *Journal of Computational Intelligence in Finance*, 7: 18–35, 1999.
8 O. Renaud, J.-L. Starck, and F. Murtagh. Wavelet-based combined signal filtering and prediction. *IEEE Transactions on Systems, Man and Cybernetics, Part B, Cybernetics*, 35(6): 1241–1251, 2005.
9 A.B. Geva. Scalenet multiscale neural network architecture for time series prediction. *IEEE Transactions on Neural Networks*, 9(6): 1471–1482, 1998.
10 S. Bai, J.Z. Kolter, and V. Koltun. An empirical evaluation of generic convolutional and recurrent networks for sequence modeling. *arXiv: 1803.01271*, 2018.
11 J.-S. Jang. ANFIS: Adaptive-network-based fuzzy inference system. *IEEE Transactions on Systems, Man, Cybernetics*, 23(3): 665–685, 1993.
12 S. Yang and J. Liu. Time-series forecasting based on high-order fuzzy cognitive maps and wavelet transform. *IEEE Transactions on Fuzzy Systems*, 26(6): 3391–3402, 2018.
13 Z. Liu and J. Liu. A robust time series prediction method based on empirical mode decomposition and high-order fuzzy cognitive maps. *Knowledge-Based Systems*, 203: 106105, 2020.
14 P. Liu, J. Liu, and K. Wu. CNN-FCM: systems modeling promotes stability of deep learning in time series prediction. *Knowledge-Based Systems*, 203: 106081, 2020.
15 R. Gao, L. Du, O. Duru, and K.F. Yuen. Time series forecasting based on echo state network and empirical wavelet transformation. *Applied Soft Computing*, 102: 107111, 2021.
16 S. Panigrahi and H.S. Behera. A hybrid ETS-ANN model for time series forecasting. *Engineering Applications of Artificial Intelligence*, 66: 49–59, 2017.
17 C. Luo, C. Tan, and Y. Zheng. Long-term prediction of time series based on stepwise linear division algorithm and time-variant zonary fuzzy information granules. *International Journal of Approximate Reasoning*, 108: 38–61, 2019.
18 K. Yuan, J. Liu, S. Yang, K. Wu, and F. Shen. Time series forecasting based on kernel mapping and high-order fuzzy cognitive maps. *Knowledge-based Systems*, 206: 106359, 2020.
19 S. Du, T. Li, Y. Yang, and S.-J. Horng. Multivariate time series forecasting via attention-based encoder-decoder framework. *Neurocomputing*, 388: 269–279, 2020.

20 G.E.P. Box, K.P. Lin, and D.A. Pierce. Distribution of residual autocorrelations in autoregressive-integrated moving average time series models. *Journal of the American Statistical Association*, 65(332): 1509–1520, 1970.

21 S. Yao, S. Hu, Y. Zhao, A. Zhang, and T. Abdelzaher. Deepsense: a unified deep learning framework for time-series mobile sensing data processing. *Proceedings of the 26th International Conference on World Wide Web*, pages 351–360, 2017.

22 P.F. Pai, K.P. Lin, C.S. Lin, and P.T. Chang. Time series forecasting by a seasonal support vector regression model. *Expert Systems with Applications*, 37(6): 4261–4265, 2010.

23 V. Kuznetsov and Z. Mariet. Foundations of sequence-to-sequence modeling for time series. *arXiv:1805.03714*, 2018.

24 A. Miranian and M. Abdollahzade. Developing a local least-squares support vector machines-based neuro-fuzzy model for nonlinear and chaotic time series prediction. *IEEE Transactions on Neural Networks and Learning Systems*, 24(2): 207–218, 2013.

25 H. Wang and X. Yan. Optimizing the echo state network with a binary particle swarm optimization algorithm. *Knowledge-Based Systems*, 86: 182–193, 2015.

26 R. Tibshirani. Regression shrinkage and selection via the lasso. *Journal of the Royal Statistical Society, Series B (Methodological)*, 58(1): 267–288, 1996.

27 X. Dutoit, B. Schrauwen, J. Van Campenhout, D. Stroobandt, H. Van Brussel, and M. Nuttin. Pruning and regularization in reservoir computing. *Neurocomputing*, 71(7–9): 1534–1546, 2009.

28 Y. Xue, L. Yang, and S. Haykin. Decoupled echo state networks with lateral inhibition. *Neural Networks*, 20(3): 365–376, 2007.

29 A. Rodan and P. Tino. Minimum complexity echo state network. *IEEE Transactions on Neural Networks*, 22(1): 131–144, 2011.

30 H. Jaeger and H. Haas. Harnessing nonlinearity: predicting chaotic systems and saving energy in wireless communication. *Science*, 304(5667): 78–80, 2004.

31 J. Qiao, L. Wang, C. Yang, and K. Gu. Adaptive Levenberg-Marquardt algorithm based echo state network for chaotic time series prediction. *IEEE Access*, 6: 10731, 2018.

32 D. Li, M. Han, and J. Wang. Chaotic time series prediction based on a novel robust echo state network. *IEEE Transactions on Neural Networks and Learning Systems*, 23(5): 787–799, 2012.

33 M. Han, S. Zhang, M. Xu, T. Qiu, and N. Wang. Multivariate chaotic time series online prediction based on improved kernel recursive least squares algorithm. *IEEE Transactions on Cybernetics*, 49(4): 1160–1172, 2019.

34 W.F. Liu, I. Park, Y.W. Wang, and J.C. Principe. Extended kernel recursive least squares algorithm. *IEEE Transactions on Signal Processing*, 57(10): 3801–3814, 2009.

35 S.V. Vaerenbergh, J. Via, and I. Santamaria. A sliding-window kernel RLS algorithm and its application to nonlinear channel identification. *Proceedings of the IEEE International Conference on Acoustics, Speech, and Signal Processing (ICASSP)*, pages V789–V792, 2006.

36 Y. Engel, S. Mannor, and R. Meir. The kernel recursive least-squares algorithm. *IEEE Transactions on Signal Processing*, 52(8): 2275–2285, 2004.

37 S.V. Vaerenbergh, I. Santamaría, W.F. Liu, and J.C. Principe. Fixed-budget kernel recursive least-squares. *Proceedings of the IEEE International Conference on Acoustics, Speech, and Signal Processing (ICASSP)*, pages 1882–1885, 2010.

38 S.V. Vaerenbergh, M. Lazaro-Gredilla, and I. Santamaría. Kernel recursive least-squares tracker for time-varying regression. *IEEE Transactions on Neural Networks and Learning Systems*, 23(8): 1313–1326, 2012.

39 B.D. Chen, S.L. Zhao, P.P. Zhu, and J.C. Principe. Quantized kernel recursive least squares algorithm. *IEEE Transactions on Neural Networks and Learning Systems*, 24(9): 1484–1491, 2012.

40 M.C. Mackey and L. Glass. Oscillation and chaos in physiological control systems. *Science*, 197: 287–289, 1997.

41 H.J. Song, C.Y. Miao, and Z.Q. Shen. A fuzzy neural network with fuzzy impact grades. *Neurocomputing*, 72: 3098–3122, 2009.

42 N.K. Kasabov and Q. Song. Denfis: dynamic evolving neural-fuzzy inference system and its application for time-series prediction. *IEEE Transactions on Fuzzy Systems*, 10: 144–154, 2002.

43 C. Juang and Y. Tsao. A self-evolving interval type-2 fuzzy neural network with on-line structure and parameter learning. *IEEE Transactions on Fuzzy Systems*, 16: 1411–1424, 2008.

44 H.J. Song, C.Y. Miao, Z.Q. Shen, W. Roel, D.H. Maja, and C. Francky. Design of fuzzy cognitive maps using neural networks for predicting chaotic time series. *Neural Networks*, 23: 1264–1275, 2010.

45 J.C. Wang, W.Z. Zhu, and Y.B. Zhang. Improved algorithm of grey incidence degree based on area. *Journal of Systems Engineering and Electronics*, 32(4): 777–779, 2010.

46 S.F. Liu, N.M. Xie, and J. Forrest. Novel models of grey relational analysis based on visual angle of similarity and nearness. *Grey systems - Theory and Applications*, 1(1): 8–18, 2011.

47 J.L. Deng. Introduction to grey system theory. *Journal of Grey System*, 1(1): 1–24, 1989.

48 M. Han, R. Zhang, T. Qiu, M. Xu, and W. Ren. Multivariate chaotic time series prediction based on improved grey relational analysis. *IEEE Transactions on Cybernetics*, 49(10): 2144–2154, 2019.

4

Echo-like Models for Time Series Forecasting

4.1 Introduction

The inherent recurrence in recurrent neural networks (RNNs) makes them ideal for time series analysis, though their practical use faces major hurdles due to difficulties in modeling long-term dependencies. The RNN variants represented by long short-term memory (LSTM) and gated recurrent unit (GRU) have been extensively applied in time series prediction because of the introduction of gating mechanisms, which can greatly alleviate this long-range dependency problem. The parameters of LSTM and GRU networks are commonly optimized through the backpropagation through time (BPTT) method. When the time span is large, they cannot avoid the long-range dependency problem caused by gradient vanishing.

The echo state network (ESN), a novel variant of RNNs, demonstrates remarkable efficacy in time series forecasting and analysis due to its unique reservoir computing framework and efficient training method. Its core feature lies in capturing the nonlinear characteristics of dynamic systems through a randomly initialized sparse connection reservoir, which only requires training the linear weights of the output layer, thereby reducing computational complexity. In recent years, with the explosive growth of time series data in industries, finance, meteorology, and other fields, the potential of ESN in prediction accuracy, model generalization ability, and computational efficiency has been widely explored.

In this chapter, three variants of ESN are discussed, which are constructed by nonlinear spiking neural P (NSNP) systems. These variants are called echo-like models [1–3]. A fundamental difference from existing ESN variants is that these three variants use the NSNP systems as their reservoir.

Spiking Neural P Systems for Time Series Analysis, First Edition. Jun Wang and Hong Peng.

4.2 Echo State Networks

ESN constitutes a new class of RNNs. Different from RNN variants (such as LSTM and GRU), ESN has a dynamic reservoir and recurrent-free readout. The reservoir provides a nonlinear dynamics, and the readout severs as output layer to generate the desired output from the reservoir. Consequently, echo state network achieves functional decoupling between reservoir dynamics and readout transformation, a mechanism analogous to kernel-driven approaches. The functions of state variable $v(\cdot)$ and output $y(\cdot)$ are distinct: $v(\cdot)$ transforms the historical input sequence $x(\tau), x(\tau-1), \ldots$ into a high-dimensional reservoir state $v(\tau) \in R^n$, while $y(\cdot)$ synthesizes the neuronal states $v(\tau)$ to generate target outputs. Figure 4.1 shows the structure of a typical ESN.

Given the distinct roles of the reservoir and readout components, employing separate training/generation processes with customized objective functions proves an effective strategy. The output measurement $y(\tau)$, defined as $y(x(\tau))$, is fundamentally a time-independent function, making its learning process comparatively straightforward. Unlike conventional recurrent neural networks, the reservoir's parameters are not subject to training. These connection weights are randomly assigned at initialization and kept fixed throughout the learning process, with solely the readout layer's weights being adjustable.

ESN presents an innovative approach for solving regression tasks. This section provides a concise overview of the leaky integrator variant. Consider an echo state network comprising m input nodes and n reservoir neurons.

The reservoir's status gets updated through

$$v(\tau) = (1-\beta) \cdot v(\tau-1) + \beta \cdot \tanh(W_{\text{in}}x(\tau) + Wv(\tau-1) + \theta) \tag{4.1}$$

where the reservoir's state vector at timestep τ is represented by $v(\tau) \in \mathbb{R}^n$, while the input vector is given by $x(\tau) \in \mathbb{R}^m$; the transformation from input to reservoir space is governed by weight matrix $W_{\text{in}} \in \mathbb{R}^{m\times n}$; internal reservoir dynamics are controlled by recurrent weight matrix $W \in \mathbb{R}^{n\times n}$, with $\theta \in \mathbb{R}^n$ serving as the bias vector; the system's memory retention is adjusted through leak rate parameter $\beta \in [0, 1]$.

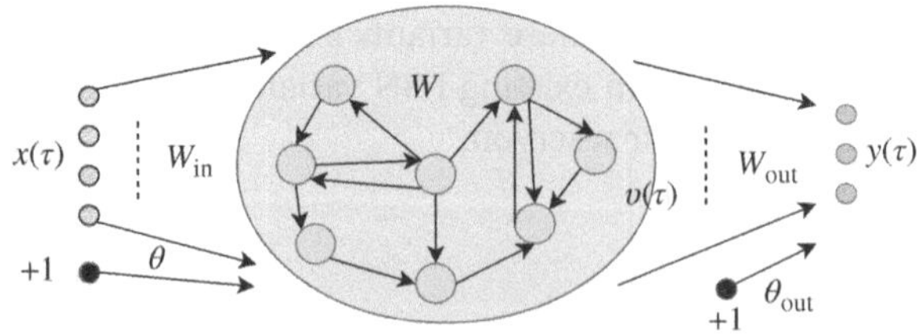

Figure 4.1 The structure of ESN (provided by authors).

Echo state network results are obtained via

$$y(\tau) = W_{\text{out}}v(\tau) + \theta_{\text{out}} \tag{4.2}$$

wherein $y(\tau)$ denotes output at timestep τ, $W_{\text{out}} \in \mathbb{R}^{k \times n}$ is the weight from reservoir to output, and $\theta_{\text{out}} \in \mathbb{R}^k$ denotes the bias.

A linear output layer is commonly employed, where its training process essentially reduces to a linear regression task. Consequently, standard linear regression techniques are directly applicable for optimizing this layer. More details can be found in Refs. [4–6].

4.3 Echo-like Spiking Neural P Model

4.3.1 A Specialized NSNP System and Its Extension

In this section, a specialized nonlinear spiking neural P (NSNP) system and its extension are introduced. The specialized NSNP system is defined, and then an extension is given.

Definition 4.1: A specialized NSNP system is defined by

$$\Pi = (O, \sigma_1, \sigma_2, \ldots, \sigma_n, syn, x)$$

wherein

1. $O = \{\alpha\}$ is an alphabet, where α is known as the spike;
2. $\sigma_1, \sigma_2, \ldots, \sigma_n$ are n neurons, with $\sigma_i = (v_i, r_i)$, $i = 1, 2, \ldots, n$, wherein
   a. $v_i \in \mathbb{R}$ is the state of σ_i.
   b. r_i is a nonlinear spiking rule $\alpha^{g(v)} \to \alpha^{f(v)}$, where $v = x + v_i$, and functions $f(\cdot)$ and $g(\cdot)$ are nonlinear.
3. $syn \subseteq \{1, \ldots, n\} \times \{1, \ldots, n\}$ with $i \neq j$, i, $j = 1, \ldots, n$ (synapses).
4. x denotes the external input.

As depicted in Figure 4.2, the customized NSNP framework contains n neuronal units $(\sigma_1, \sigma_2, \ldots, \sigma_n)$ that receive x as their input signal. In the specialized NSNP system, every neuron contains a nonlinear spiking rule $\alpha^{g(v)} \to \alpha^{f(v)}$, an external input x and a state unit v_i. Within the spiking rule framework, the term $g(\cdot)$ denotes the consumption function, while $f(\cdot)$ represents the generation function.

Let $v_i(\tau)$ represent the activation state of σ_i at timestep τ, while $x(\tau)$ corresponds to the external stimulus. Upon activation of σ_i, it utilizes spikes valued at $g(v)$ while producing new spikes with a value of $f(v)$, where $v = x(\tau) + v_i(\tau - 1)$. The

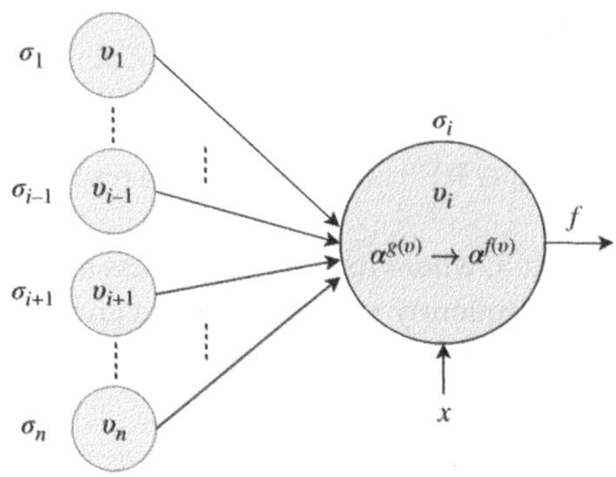

Figure 4.2 The specialized NSNP system with external input [1]/with permission of Elsevier.

produced spikes are transmitted to its post-synaptic neurons. Thus, based on non-linear spiking mechanism, state equation for σ_i is given by

$$v_i(\tau) = v_i(\tau - 1) - g(x(\tau) + v_i(\tau - 1)) + \sum_{j=1, j \neq i}^{n} f(x(\tau) + v_j(\tau - 1)), \tag{4.3}$$

wherein $i = 1, 2, \ldots, n$. Assume that the specialized NSNP system has m external inputs $x_1, x_2, \ldots, x_m$, and let $x = \sum_{s=1}^{m} x_s$. Therefore, we obtain

$$v_i(\tau) = v_i(\tau - 1) - g\left(\sum_{s=1}^{m} x_s(\tau) + v_i(\tau - 1)\right) + \sum_{j=1, j \neq i}^{n} f\left(\sum_{s=1}^{m} x_s(\tau) + v_j(\tau - 1)\right) \tag{4.4}$$

Let $g(\cdot) \equiv -f(\cdot)$. The state of neuron σ_i is computed by

$$v_i(\tau) = v_i(\tau - 1) + \sum_{j=1}^{n} f\left(\sum_{s=1}^{m} x_s(\tau) + v_j(\tau - 1)\right) \tag{4.5}$$

Then, the specialized NSNP system is parameterized. Therefore, state of σ_i can be updated as follows:

$$v_i(\tau) = \beta_1 \cdot v_i(\tau - 1) + \sum_{j=1}^{n} w_{ij}^1 \cdot f\left(\sum_{s=1}^{m} w_{sj}^2 \cdot x_s(\tau) + \beta_2 \cdot v_j(\tau - 1)\right), \tag{4.6}$$

where β_1, β_2, w_{ij}^1, and w_{sj}^2 are the parameters.

The NSNP framework is additionally characterized as a MIMO (multi-input, multi-output) architecture. Suppose that $\mathbf{v}(\tau) = [v_1(\tau), v_2(\tau), \ldots, v_n(\tau)]^T$ and $\mathbf{x}(\tau) = [(x_1(\tau), \ldots, x_m(\tau)]^T$. Thus, the system's state is updated as follows:

$$\mathbf{v}(\tau) = \beta_1 \cdot \mathbf{v}(\tau - 1) + Wf(W_{in}\mathbf{x}(\tau) + \beta_2 \cdot \mathbf{v}(\tau - 1)), \tag{4.7}$$

where β_1 and β_2 are scalar factors and $W = (w_{ij}^1)_{n\times n}$ and $W_{\text{in}} = (w_{sj}^2)_{m\times n}$ are two parameter matrices.

4.3.2 ESNP Model

Structurally, an ESN consists of three key elements: a receiving layer for input signals, a dynamic processing layer (referred to as the reservoir) with interconnected neural units, and a final projection layer (designated as the readout).

Building upon ESN's topological architecture, an innovative recurrent-type computational model is introduced, named echo-like spiking neural P (ESNP) systems or simply ESNP model [1]. The structural framework of ESNP systems comprises three elements – an input layer, reservoir, and output layer. A key distinction emerges in the reservoir composition: NSNP implementations feature spiking neurons, whereas ESN adopts standard neuron models. In other words, the reservoir is actually a specialized NSNP system.

Figure 4.3 illustrates the ESNP model (or ESNP system). It consists of three elements: (i) The input layer contains m inputs $\mathbf{x}(\tau) = [x_1(\tau), \ldots, x_m(\tau)]^T$; (ii) the reservoir consists of n interconnected neurons forming an internal network; (iii) this architecture employs k neurons in its final layer, analogous to ESN's readout mechanism.

Let $W_{\text{in}} = (w_{ij}^2)_{m\times n} \in \mathbb{R}^{m\times n}$ be the matrix from input to reservoir, $W = (w_{ij}^1)_{n\times n} \in \mathbb{R}^{n\times n}$ denote the interconnected matrix in the reservoir, $W_{\text{out}} = (w_{ij}^3)_{n\times k} \in \mathbb{R}^{n\times k}$ be the reservoir-output matrix, and $W'_{\text{out}} = (w_{ij}^4)_{m\times k} \in \mathbb{R}^{m\times k}$ be the input-output matrix.

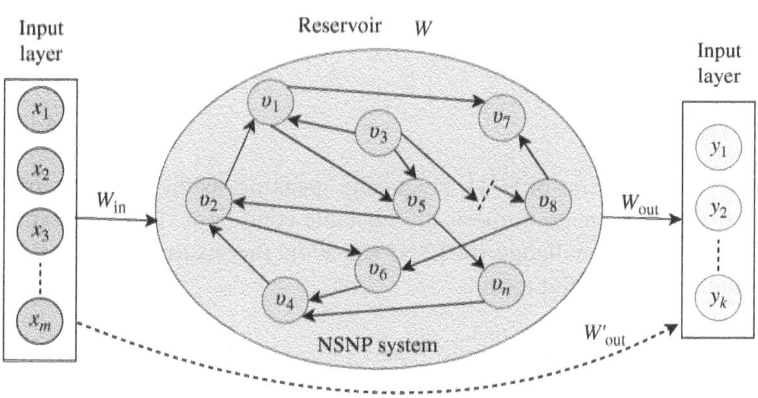

Figure 4.3 Echo spiking neural P system or ESNP model [1]/with permission of Elsevier.

The ESNP system's update rules are derived from Eq. (4.7), expressed as

$$\mathbf{v}(\tau) = \beta_1 \cdot \mathbf{v}(\tau - 1) + Wf(W_{\text{in}}\mathbf{x}(\tau) + \beta_2 \cdot \mathbf{v}(\tau - 1)) \tag{4.8}$$

$$\mathbf{y}(\tau) = W'_{\text{out}}\mathbf{x}(\tau) + W_{\text{out}}\mathbf{v}(\tau) \tag{4.9}$$

wherein $\mathbf{y}(\tau) = [y_1(\tau), \ldots, y_k(\tau)]^T$ are the outputs at time τ, $\beta_1, \beta_2 \in [0, 1]$ are two scalar factors, $W_{\text{in}} \in \mathbb{R}^{m\times n}$, $W \in \mathbb{R}^{n\times n}$, $W_{\text{out}} \in \mathbb{R}^{n\times k}$, and $W'_{\text{out}} \in \mathbb{R}^{m\times k}$ are four weight parameters. Stated alternatively, $f(\cdot)$ belongs to the sigmoid family, where $\tanh(\cdot)$ serves as a representative case.

Given the architectural parallels between ESN and ESNP, the latter can adopt comparable supervised learning methods for system training. Mirroring ESN's approach, ESNP's learning process consists of two phases: (i) network initialization and (ii) ESNP training.

At the beginning, matrices W_{in} and W are randomly initialized and fixed. At training stage, ESNP aims to optimize the matrices W_{out} and W'_{out}.

Suppose that $\mathbf{x}(\tau) = [x_1(\tau), \ldots, x_m(\tau)]^T \in \mathbb{R}^m$ denotes the input, and $\hat{\mathbf{y}}(\tau) = [\hat{y}_1(\tau), \ldots, \hat{y}_k(\tau)]^T \in \mathbb{R}^k$ denotes the estimated output. Let $\mathbf{W}_{\text{out}} = [W^T_{\text{out}}, W'^T_{\text{out}}] \in \mathbb{R}^{k\times(n+m)}$ and $\mathbf{X}(\tau) = [\mathbf{v}(\tau)^T, \mathbf{x}(\tau)^T]^T \in \mathbb{R}^{n+m}$. Therefore, $\mathbf{y}(\tau)$ is calculated by

$$\mathbf{y}(\tau) = \mathbf{W}_{\text{out}}\mathbf{X}(\tau) \tag{4.10}$$

Assume that $\mathbf{x}(1), \ldots, \mathbf{x}(L)$ are L available data and $\hat{\mathbf{y}}(1), \ldots, \hat{\mathbf{y}}(L)$ are the estimated outputs. Let $\mathbf{Y} = [\hat{\mathbf{y}}(1), \ldots, \hat{\mathbf{y}}(L)]$ and $\mathbf{H} = [\mathbf{X}(1), \ldots, \mathbf{X}(L)]$. Therefore, we can obtain

$$\mathbf{Y} = \mathbf{W}_{\text{out}}\mathbf{H} \tag{4.11}$$

Then, $\mathbf{W}_{\text{out}}$ can be computed by

$$\mathbf{W}_{\text{out}} = (\mathbf{H}^T\mathbf{H})^{-1}\mathbf{H}\mathbf{Y}, \tag{4.12}$$

where $(\mathbf{H}^T\mathbf{H})^{-1}$ is the inverse of $\mathbf{H}^T\mathbf{H}$. Practically speaking, as the size of matrix $\mathbf{H}$ increases, it tends to create numerically unstable conditions. For addressing this numerically unstable challenge, the ESNP can adopt the ensuing optimization function by adding a constraint term:

$$J(\mathbf{W}_{\text{out}}) = \frac{1}{2}||\mathbf{Y} - \mathbf{W}_{\text{out}}\mathbf{H}||^2 + \frac{1}{2}\lambda||\mathbf{W}_{\text{out}}||^2, \tag{4.13}$$

wherein λ denotes a regularization factor. The output matrix $\mathbf{W}_{\text{out}}$ is determined by

$$\mathbf{W}_{\text{out}} = (\mathbf{H}^T\mathbf{H} + \lambda\mathbf{I})^{-1}\mathbf{H}\mathbf{Y}, \tag{4.14}$$

where $\mathbf{I} \in \mathbb{R}^{n \times n}$ denotes the identity matrix. When executing the ESNP framework, the weight matrix $\mathbf{W}_{\text{out}}$ may be determined through standard ridge regression technique:

$$\mathbf{W}^*_{\text{out}} = \arg\min_{\mathbf{W}_{\text{out}}} \left\{ J(\mathbf{W}_{\text{out}}) = \frac{1}{2}||\mathbf{Y} - \mathbf{W}_{\text{out}}\mathbf{H}||^2 + \frac{1}{2}\lambda||\mathbf{W}_{\text{out}}||^2 \right\} \tag{4.15}$$

The key distinctions between ESNP systems and both SNP/ESN systems are as below:

1. ESNP systems represent a distinct nonlinear adaptation of standard SNP systems, characterized by their incorporation of both input and readout layers as specialized NSNP architectures. This structural modification fundamentally distinguishes them from conventional SNP systems.
2. ESNP systems represent a modified form of ESN, distinguished primarily by employing an NSNP-based reservoir structure and incorporating distinct update mechanisms compared to traditional ESNs.

4.3.3 Model Evaluation

This section evaluates the ESNP model's effectiveness through time series prediction experiments. The investigation involves three distinct time series datasets, with performance comparisons made against 10 state-of-the-art algorithms and 17 fundamental forecasting methods. The experimental datasets include:

1. Six benchmark datasets: Lake, CO_2, Critical, Closing, SP500, and MG;
2. Three larger real-life datasets: Sunspot, Beijing PM2.5, and Brent oil price datasets;
3. Two synthetic datasets: Rossler chaotic and Lorenz chaotic datasets.

As shown in Table 4.1, an overview of the three time series groups is presented, where each series' characteristics have been normalized within the $[-1, 1]$.

The ESNP architecture incorporates six hyperparameters: (i) A pair of scaling coefficients β_1 and β_2 defined in Eq. (4.8); (ii) Network dimension n, representing neuron count in the NSNP framework; (iii) Initial configuration parameters including spectral radius ρ and connection sparsity δ; (iv) The regularization coefficient λ. Table 4.2 presents the fine-tuned hyperparameter configurations, where each parameter set is systematically determined through individual time series grid optimization.

Table 4.1 Statistical information of time series [1]/with permission of Elsevier.

Name	Size	Variables number	Initial size	Training size	Test size
MG	1,000	1	50	450	500
SP500	251	1	30	120	101
Closing	291	1	10	208	73
Critical	240	1	10	210	20
CO_2	192	1	10	153	29
Lake	590	1	25	425	140
Sunspot	3,132	1	200	1,900	1,032
Beijing PM2.5	43,800	9	1,000	34,040	8,760
Brent oil price	8,216	1	400	6,994	822
Lorenz	3,000	3	500	1,900	600
Rossler	5,500	3	300	4,290	910

Table 4.2 Hyperparameter setting of the ESNP model [1]/with permission of Elsevier.

Name	λ	δ	ρ	m	β_1	β_2
Rossler	1.0e-06	40%	0.85	59	0.02	0.63
Lorenz	1.0e-06	40%	0.98	146	0.03	0.29
Sunspot	1.0e-06	40%	0.98	17	0.02	0.75
Beijing PM2.5	1.0e-06	40%	0.98	51	0.43	0.16
Brent oil price	1.0e-06	40%	0.98	22	0.30	0.15
Lake	1.0e-06	40%	0.98	22	0.70	1.00
CO_2	1.0e-06	40%	0.98	23	0.38	0.66
Critical	1.0e-06	40%	0.98	27	0.70	0.39
Closing	1.0e-06	40%	0.98	11	0.18	0.34
SP500	1.0e-06	40%	0.98	28	0.03	0.31
MG	1.0e-06	40%	0.98	83	0.02	0.32

To quantitatively analyze the prediction models' performance, a pair of evaluation criteria are utilized: RMSE (root mean square error) and MAE (mean absolute error) indices. Their formal definitions can be written as:

$$\mathrm{RMSE} = \sqrt{\frac{1}{T}\sum_{\tau=1}^{T}(x(\tau) - y(\tau))^2} \tag{4.16}$$

$$\text{MAE} = \frac{1}{T} \sum_{\tau=1}^{T} |x(\tau) - y(\tau)| \tag{4.17}$$

At each time instance τ, $x(\tau)$ corresponds to the observed value and $y(\tau)$ to its estimated counterpart, with T standing for the sample size. In both evaluation metrics, lower numerical values typically correspond to superior predictive accuracy.

4.3.3.1 Six Benchmark Datasets

The first group dataset comprises six short univariate temporal sequences: Lake, CO_2, Critical, Closing, SP500, and MG. The performance of the ESNP framework is assessed through six standard temporal datasets and benchmarked against six leading forecasting approaches.

Table 4.3 gives the comparisons of ESNP model with the six prediction methods on the six datasets in terms of RMSE index. The results of LSTM-SNP and GSNP, respectively, are obtained from Refs. [7] and [8]. The results of HFCM and TCN, respectively, are obtained from Refs. [9] and [10]. The results of CNN-FCM and ANFIS, respectively, are obtained from Refs. [11] and [12]. Table 4.3 reveals the following key findings: (i) The ESNP model demonstrates superior predictive accuracy across five datasets: Lake, CO_2, Critical, Closing, and MG; (ii) regarding SP500 predictions, GSNP outperforms other models, followed by ANFIS in second place, while ESNP ranks third. The experimental findings demonstrate the ESNP model's superior performance in forecasting single-variable temporal sequences.

4.3.3.2 Three Larger Real-life Datasets

The second experimental phase evaluates the ESNP model's forecasting capabilities across three major real-world datasets: historical Sunspot observations, Beijing's multivariate PM2.5 concentrations, and Brent crude oil prices.

For the Brent oil price time series, the ESNP model is compared with five baseline and a state-of-the-art forecasting methods. Table 4.4 gives the comparison of ESNP model with six methods on MAE index. The results of the six methods are obtained from Ref. [13]. Table 4.4 shows that ESNP model outperforms the

Table 4.3 The RMSE results of the distinct methods on eight univariate time series [1]/with permission of Elsevier.

Models	Lake	CO_2	Critical	Closing	SP500	MG
ANFIS	0.4580	0.9010	0.6510	27.5620	14.9350	0.0010
CNN-FCM	0.3913	0.7305	0.5665	25.1895	20.8158	0.0013
TCN	0.4093	3.1202	0.6020	25.2143	51.2674	0.0008
HFCM	0.3770	0.5600	0.5470	23.1590	16.1050	0.0040
GSNP	0.4605	0.4248	0.6882	24.3763	11.2057	0.0013
LSTM-SNP	0.6380	—	—	24.4279	17.1704	0.0042
ESNP	0.3663	0.3855	0.4002	22.4705	16.0838	0.0002

Table 4.4 The results of the different methods on Brent dataset [1]/with permission of Elsevier.

Model	MAE
ShallowPIPE	1.18
DeepMVE	1.21
GBDT	1.38
SVR	1.23
ARIMA	1.14
DeepPIPE	1.02
ESNP	0.65

six methods. The ESNP model demonstrates superior performance in forecasting Brent crude oil price trends compared to other methods.

In evaluating Beijing's PM2.5 concentration trends, the ESNP framework undergoes performance benchmarking against one state-of-the-art and six conventional forecasting approaches. The RMSE and MAE performance metrics of the ESNP model and seven compared models are contrasted in Table 4.5. The results of the seven compared models are retrieved from Ref. [14]. Table 4.5 shows that ESNP model outperforms the seven methods. The comparisons demonstrate the advantage of ESNP model to predict the Beijing PM2.5.

For the Sunspot dataset, ESNP model is compared with the two forecasting methods. Table 4.6 displays the RMSE metric evaluation results, contrasting the ESNP approach with two alternative models whose outcomes are sourced from Ref. [6]. The comparative analysis presented in Table 4.6 reveals the ESNP method's slight advantage over Anti-Oja and its enhanced prediction accuracy relative to BCM, indicating its appropriateness for Sunspot temporal sequence forecasting.

4.3.3.3 Two Synthetic Time Series

The third experimental phase evaluates the ESNP model's forecasting capabilities using two nonlinear dynamical systems: the multivariate Lorenz and Rössler attractor datasets.

The RMSE performance of the ESNP framework is evaluated against five benchmark models and the state-of-the-art method for two chaotic time series, as presented in Table 4.7. The performance metrics of the six benchmarked models originate from Ref. [15]. Table 4.7 demonstrates that: (i) Regarding the Lorenz time series, ESNP shows marginally superior performance to mmCesN while

Table 4.5 The results of the different methods on Beijing PM2.5 dataset [1]/with permission of Elsevier.

Model	MAE	RMSE
ESNP	0.0147	0.0474
MsA+AdaHMG+TG-LSTM	0.1388	0.2723
MsA+AdaHMG	0.1445	0.2781
MsA	0.1454	0.2794
Attention-RNN	0.4349	0.7504
DA-RNN	0.1592	0.3031
LSTM	0.1638	0.3310
ARIMA-GARCH	0.5678	0.8323

Table 4.6 The results of the different models on Sunspot dataset [1]/with permission of Elsevier.

Model	RMSE
BCM	0.0441
Anti-Oja	0.0161
ESNP	0.0104

Table 4.7 The RMSE values of the different models [1]/with permission of Elsevier.

Model	Rossler	Lorenz
PSS	3.134e-1	4.743e-1
BP	7.536e-4	1.081e-3
Elman	1.412e-2	2.543e-2
ESN	5.034e-4	9.419e-4
nCESN	5.887e-3	5.681e-3
mmCESN	1.446e-4	2.668e-4
ESNP	2.150e-6	1.026e-4

significantly exceeding the predictive capabilities of the remaining five models; (ii) In the case of Rossler time series, ESNP exhibits the highest performance among all benchmarked models. The comparisons indicate that ESNP model has a significant advantage to predict the two chaotic datasets.

4.4 Echo-like Feedback Spiking Neural P Model

4.4.1 Nonlinear Spiking Neural P Systems with Two Outputs

A variant of NSNP systems [16] is introduced in this section, called nonlinear spiking neural P system with two outputs or NSNP-TO system. The NSNP-TO system is first defined, and then its mathematical model is given.

Definition 4.2: A NSNP-TO system can be given as

$$\Pi = (O, \sigma_1, \sigma_2, \ldots, \sigma_n, syn, x)$$

where

1. $O = \{\alpha\}$ is the singleton alphabet, wherein α is known as the spike;
2. $\sigma_1, \sigma_2, \ldots, \sigma_n$ are n neurons, of the form $\sigma_i = (v_i, r_i)$, $i = 1, 2, \ldots, n$, where
   a. $v_i \in \mathbb{R}$ denotes the state of σ_i;
   b. r_i indicates nonlinear spiking rule $\alpha^{g(v)} \rightarrow \alpha^{f(v)}\alpha^{f_o(v)}$, in which $v = x + v_i$, and functions $g(\cdot)$, $f(\cdot)$, and $f_o(\cdot)$ are nonlinear;
3. $syn \subseteq \{1, 2, \ldots, n\} \times \{1, 2, \ldots, n\}$ with $i \neq j$, $i, j = 1, 2, \ldots, n$ (synapses);
4. x is the external input.

Figure 4.4 depicts a NSNP-TO system with n neurons, $\sigma_1, \sigma_2, \ldots, \sigma_n$. In the NSNP-TO system, every neuron contains a state unit v_i and nonlinear spiking rule $\alpha^{g(v)} \rightarrow \alpha^{f(v)}\alpha^{f_o(v)}$. Assume that $v_i(\tau)$ is the state of σ_i at timestep τ. If neuron σ_i fires, it removes the spikes of $g(v)$, and then produces two outputs: (i) the spikes with value $f(v)$, sending to its post-synaptic neurons; (ii) the spikes with value $f_o(v)$,

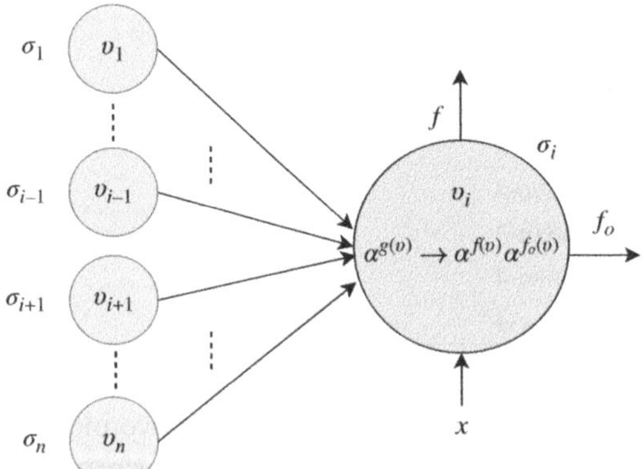

Figure 4.4 A NSNP-TO system [2]/with permission of Elsevier.

regarded as its external output, in which $v = x(\tau) + v_i(\tau - 1)$, $x(\tau)$ is the current input, and $v_i(\tau - 1)$ is the state at time $\tau - 1$.

The state updating for σ_i is given by

$$v_i(\tau) = v_i(\tau - 1) - g(x(\tau) + v_i(\tau - 1)) + m_i \tag{4.18}$$

where m_i is the spikes received by neuron σ_i from its presynaptic neurons, and is given by

$$m_i = \sum_{j=1, j \neq i}^{n} f(x(\tau) + v_j(\tau - 1)) \tag{4.19}$$

wherein $i = 1, \ldots, n$.

The output of σ_i is computed by

$$h_i(\tau) = f_o(x(\tau) + v_i(\tau - 1)) \tag{4.20}$$

For simplified model formulation, define the function as $g(v) = -f(v)$. Therefore, we have

$$v_i(\tau) = v_i(\tau - 1) + \sum_{j=1}^{n} f(x(\tau) + v_j(\tau - 1)) \tag{4.21}$$

Let $x = \sum_{s=1}^{m} x_s$. The state of σ_i is updated by

$$v_i(\tau) = v_i(\tau - 1) + \sum_{j=1}^{n} f\left(\sum_{s=1}^{m} x_s(\tau) + v_j(\tau - 1)\right) \tag{4.22}$$

The output of σ_i is computed as

$$h_i(\tau) = f_o\left(\sum_{s=1}^{m} x_s(\tau) + v_i(\tau - 1)\right) \tag{4.23}$$

The NSNP-TO system is further parameterized, and the state equation for σ_i is given by

$$v_i(\tau) = \beta_1 \cdot v_i(\tau - 1) + \sum_{j=1}^{n} w_{ij}^1 \cdot f\left(\sum_{s=1}^{m} w_{is}^2 \cdot x_s(\tau) + \beta_2 \cdot v_j(\tau - 1)\right) \tag{4.24}$$

where $\beta_1, \beta_2, w_{ij}^1$, and w_{is}^2 are the parameters, and $i = 1, 2, \ldots, n$. The output of neuron σ_i is computed by

$$h_i(\tau) = f_o\left(\sum_{s=1}^{m} w_{is}^3 \cdot x_s(\tau) + \gamma \cdot v_i(\tau - 1)\right) \tag{4.25}$$

where γ and w_{is}^3 are the parameters.

Let $\mathbf{v}(\tau) = [v_1(\tau), \ldots, v_n(\tau)]^T$, $\mathbf{x}(\tau) = [(x_1(\tau), \ldots, x_m(\tau)]^T$, and $\mathbf{y}(\tau) = [(y_1(\tau), \ldots, y_k(\tau)]^T$. The system's state is computed by

$$\mathbf{v}(\tau) = \beta_1 \cdot \mathbf{v}(\tau - 1) + Wf\big(W_{\text{in}}\mathbf{x}(\tau) + \beta_2 \cdot \mathbf{v}(\tau - 1)\big) \tag{4.26}$$

where β_1 and β_2 are scalar factors and $W = (w_{ij}^1)_{n\times n}$ and $W_{\text{in}} = (w_{ij}^2)_{m\times n}$ are parameter matrices. Therefore, the system's output is computed as

$$\mathbf{h}(\tau) = f_o\big(W'_{\text{in}}\mathbf{x}(\tau) + \gamma \cdot \mathbf{v}(\tau - 1)\big) \tag{4.27}$$

where γ is a scalar factor, and $W'_{\text{in}} = (w_{ij}^3)_{m\times n}$ is a parameter matrix.

4.4.2 EFSNP Model

Differing from NSNP systems, NSNP-TO systems have nonlinear spiking mechanisms with two outputs. Under the control of the nonlinear spiking mechanisms, NSNP-TO systems can show dynamic behavior. Consequently, dynamic systems can be effectively modeled using NSNP-TO systems.

The ESN architecture fundamentally consists of three components: an input layer, a dynamic reservoir, and a readout layer. Notably, the reservoir component can theoretically be implemented using various dynamical systems. Consequently, the NSNP-TO framework serves as the foundational architecture for developing EFSNP, an enhanced echo state network derivative [2]. The EFSNP architecture comprises three key components: an input processing unit, a dynamic reservoir based on NSNP-TO technology, and a final readout layer for output generation. The EFSNP framework, shown in Figure 4.5, is composed of three structural elements: m input nodes forming the reception layer, a nonlinear dynamical system (NSNP-TO) with n neural units, and a decoding layer featuring k output elements. In addition, the output feedback is designed in EFSNP model.

According to Eqs. (4.26) and (4.27), the state equation and output equation are given by

$$\mathbf{v}(\tau) = \beta_1 \cdot \mathbf{v}(\tau - 1) + Wf\big(W_{\text{in}}\mathbf{x}(\tau) + \beta_2 \cdot \mathbf{v}(\tau - 1) + W_{\text{back}}\mathbf{y}(\tau - 1)\big) \tag{4.28}$$

$$\mathbf{h}(\tau) = f_o\big(W'_{\text{in}}\mathbf{x}(\tau) + \gamma \cdot \mathbf{v}(\tau - 1) + W_{\text{back}}\mathbf{y}(\tau - 1)\big) \tag{4.29}$$

$$\mathbf{y}(\tau) = W_{\text{out}}\mathbf{h}(\tau) \tag{4.30}$$

where $\mathbf{x}(\tau) \in \mathbb{R}^m$, $\mathbf{v}(\tau) \in \mathbb{R}^n$, and $\mathbf{y}(\tau) \in \mathbb{R}^k$; $W_{\text{out}} \in \mathbb{R}^{k\times n}$ is the matrix form reservoir to output; $W_{\text{back}} \in \mathbb{R}^{n\times k}$ is feedback matrix; $W \in \mathbb{R}^{n\times n}$ is the reservoir's matrix; $W'_{\text{in}} \in \mathbb{R}^{m\times n}$ is the matrix from input to output; $W_{\text{in}} \in \mathbb{R}^{m\times n}$ is the matrix from input to reservoir; β_1, β_2, and γ are the scalar factors; $f(\cdot)$ and $f_o(\cdot)$ are nonlinear functions. The implementation adopts tanh$(\cdot)$ for $f(\cdot)$, while $f_o(\cdot)$ employs the standard sigmoid function.

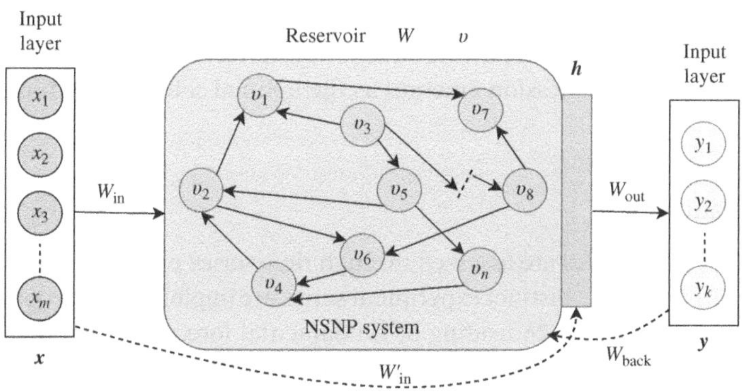

Figure 4.5 An EFSNP model, in which the reservoir is a NSNP-TO system [2]/with permission of Elsevier.

The diagram in Figure 4.5 demonstrates that the EFSNP structure can be interpreted as a modified NSNP-TO mechanism incorporating specialized input and readout components. The NSNP-TO framework maintains an internal vector $\mathbf{v}(\tau)$ while producing an output signal $\mathbf{h}(\tau)$ at time instant τ, as previously outlined. The state updating and input-output relationships of the NSNP-TO system/reservoir are mathematically formulated in Eqs. (4.28) and (4.29). The EFSNP model adopts a straightforward linear readout approach. In contrast, the EFSNP model's output (Eq. [4.30]) processes the state vector $\mathbf{h}(\tau)$ through a linear transformation.

As is typical in the ESN, the learning of the EFSNP comprises two stages: (i) the initialization stage and (ii) training stage. In the initialization stage, four matrices W_{in}, W'_{in}, W, and W_{back} are randomly initialized and remain unchanged during the training stage. Notably, γ along with β_1 and β_2 in Eqs. (4.28) and (4.29) function as predetermined hyperparameters. The training process exclusively adjusts the readout layer's weight matrix W_{out} post-initialization.

Following Eq.(4.30), the EFSNP model's training process can be formulated as a linear system solution for the output weights W_{out}:

$$W_{out}H = Y_{target} \tag{4.31}$$

wherein $\mathbf{h}(\tau)$ is the reservoir output; $H = (\mathbf{h}(1), \ldots, \mathbf{h}(T)) \in \mathbb{R}^{n\times T}$; $Y_{target} = (\mathbf{y}_{target}(1), \mathbf{y}_{target}(2), \ldots, \mathbf{y}_{target}(T)) \in \mathbb{R}^{k\times T}$, and $\mathbf{y}_{target}(\tau)$ is the estimated output.

One straightforward approach to address Eq. (4.31) is by utilizing the Moore–Penrose generalized inverse H^+ of H; thereby $W_{out} = Y_{target}H^+$. To circumvent this ill-condition issue, an objective function incorporating a regularization component is implemented as follows:

$$J(W_{out}) = \frac{1}{2}(||Y - W_{out}H||^2 + \lambda||W_{out}||^2) \tag{4.32}$$

wherein λ denotes the regularization factor. Given that Eq. (4.31) represents a linear regression model while Eq. (4.32) incorporates a regularization component in its objective function, ridge regression emerges as the optimal solution for minimizing Eq. (4.32).

4.4.3 Model Evaluation

The EFSNP model's capabilities are assessed through time series prediction tasks in experimental settings. Three distinct experiment series are implemented to contrast the EFSNP algorithm with 26 leading or fundamental forecasting methodologies. Experiments are divided into the following three groups:

1. The first group: time series prediction tasks covering both Lorenz's butterfly attractor and Rossler's spiral-chaos patterns.;
2. The second group: eight canonical datasets are analyzed, involving Lake, CO_2, Critical, Closing, Milk, SP500, MG, and Sunspot;
3. The third group: two larger scale real-life datasets: Beijing PM2.5 and Brent oil price.

Table 4.8 summarizes the three groups of time series. The features of each time series are normalized to the interval $[-1, 1]$.

Table 4.8 Statistical information of datasets [2]/with permission of Elsevier.

Name	Variables number	Size	Test size	Training size
Beijing PM2.5	9	43,800	8,760	35,040
Brent oil price	1	8,216	1,022	7,194
Lake	1	600	140	460
CO_2	1	192	29	163
Critical	1	240	20	220
Closing	1	291	73	218
Milk	1	268	34	134
SP500	1	251	101	150
MG	1	1,000	500	500
Sunspot	1	288	67	221
Rossler	3	5,500	910	4,590
Lorenz	3	3,000	600	2,400

Table 4.9 The RMSE values of the different models [2]/with permission of Elsevier.

Models	Rossler	Lorenz
PSS	3.134e-1	4.743e-1
BP	7.536e-4	1.081e-3
Elman	1.412e-2	2.543e-2
ESN	5.034e-4	9.419e-4
nCESN	5.887e-3	5.681e-3
mmCESN	1.e-4	2.668e-4
EFSNP	1.235e-4	1.187e-4

4.4.3.1 Two Chaotic Datasets

Both Lorenz and Rossler systems represent multidimensional chaotic dynamical systems characterized by time-dependent variables. Consequently, the first experimental series aims to assess the EFSNP model's capability in predicting chaotic time series.

The performance evaluation presented in Table 4.9 demonstrates the RMSE comparison between the EFSNP approach and six competing methods (including one state-of-the-art model and five baselines) for Lorenz and Rossler systems prediction. The performance results of the six benchmarked models originate from Ref. [15], with detailed configuration parameters available in cited publications. The experimental comparison presented in Table 4.9 illustrates that: (i) For Lorenz chaotic time series prediction, EFSNP maintains a small but clear lead over mmCESN while substantially outperforming five competing models; (ii) In Rossler chaotic time series forecasting, EFSNP exhibits excellent performance against all reference models. These results conclusively establish EFSNP's predictive advantage for both chaotic time series scenarios.

4.4.3.2 Eight Benchmark Datasets

Eight benchmark univariate time series are used in the second group of experiments: Sunspot, MG, SP500, Milk, Closing, Critical, CO_2, and Lake. The performance of the EFSNP framework is assessed through eight standard time series datasets and benchmarked against six state-of-the-art forecasting approaches.

Table 4.10 presents a comparative analysis of the EFSNP model against six alternative approaches across eight benchmark time series, evaluated using RMSE. The results for LSTM-SNP are obtained from Ref. [7], and the results of GSNP are obtained from Ref. [8]. HFCM data originates from the work of Ref. [9], while TCN findings are derived from Ref. [10]. The CNN-FCM outcome data originates from Ref. [11], while the ANFIS performance metrics derive from Ref. [12]. Detailed configurations of the benchmarked models are documented in their

Table 4.10 The RMSE values of the different models [2]/with permission of Elsevier.

Dataset	EFSNP	LSTM-SNP	GSNP	HFCM	TCN	CNN-FCM	ANFIS
Sunspot	16.7647	—	22.8846	18.9160	51.2674	17.9488	22.7530
MG	0.0006	0.0042	0.0013	0.0040	0.0008	0.0013	0.0010
SP500	11.8764	17.1704	11.2057	16.1050	51.2674	20.8158	14.9350
Milk	8.7741	48.3005	38.5390	8.2580	33.8579	30.4738	9.5780
Closing	21.8325	24.4279	24.3763	23.1590	25.2143	25.1895	27.5620
Critical	0.2763	—	0.6882	0.5470	0.6020	0.5665	0.6510
CO_2	0.3631	—	0.4248	0.5600	3.1202	0.7305	0.9010
Lake	0.3527	0.6380	0.4605	0.3770	0.4093	0.3913	0.4580

respective source publications. Table 4.11 presents these key findings: (i) EFSNP emerges as the top-performing model for six datasets: Lake, CO_2, Critical, Closing, MG, and Sunspot; (ii) SP500 predictions show GSNP's dominance, while EFSNP ranks second; (iii) The Milk dataset exhibits HFCM's superior performance, trailed by EFSNP. The experimental findings demonstrate that the EFSNP framework outperforms other methods in forecasting univariate time series.

4.4.3.3 Two Larger Scale Real-life Datasets

Brent crude oil prices and Beijing's PM2.5 levels represent two prominent real-world time series datasets frequently analyzed at scale. The time series data for Brent crude exhibits univariate characteristics across 8,216 observations, in contrast to Beijing's PM2.5 records, which present multivariate features with 43,800 logged measurements. The EFSNP model's capability for large-scale time series forecasting is evaluated using dual temporal datasets in the third experimental phase.

The predictive accuracy of the EFSNP algorithm is assessed relative to six competing models (including one cutting-edge technique) using Brent crude oil price data. Comparative MAE measurements appear in Table 4.11. The results of the NSNP+Wavelet model are retrieved from Ref. [17], and the results of the other six models are obtained from Ref. [13]. As demonstrated in Table 4.11, the EFSNP model exhibits superior performance compared to the six benchmark models, highlighting its exceptional capability in forecasting Brent crude oil price trends.

The EFSNP model's performance for Beijing PM2.5 forecasting is evaluated against seven benchmark methods, including one state-of-the-art approach. Comparative results of all eight models measured by RMSE and MAE metrics are presented in Table 4.12. The NSNP+Wavelet outcome data originates from Ref. [17], while the ANFIS performance metrics derive from Ref. [14]. From Table 4.12,

Table 4.11 Results of the different methods on Brent dataset [2]/with permission of Elsevier.

Model	MAE
ShallowPIPE	1.18
DeepMVE	1.21
GBDT	1.38
SVR	1.23
ARIMA	1.14
DeepPIPE	1.02
NSNP+Wavelet	1.06
EFSNP	0.63

Table 4.12 Results of the different methods on Beijing PM2.5 dataset [2]/with permission of Elsevier.

Model	MAE	RMSE
EFSNP	0.0154	0.0485
NSNP+Wavelet	0.0461	0.0748
MsA+AdaHMG+TG-LSTM	0.1388	0.2723
MsA+AdaHMG	0.1445	0.2781
MsA	0.1454	0.2794
Attention-RNN	0.4349	0.7504
DA-RNN	0.1592	0.3031
LSTM	0.1638	0.3310
ARIMA-GARCH	0.5678	0.8323

it can be seen that the EFSNP model outperforms the seven comparison methods. These comparisons illustrate the obvious advantage of the EFSNP model to predict Beijing PM2.5.

4.5 Deep Echo-like Spiking Neural P Model

4.5.1 Shallow ESNP Model

The ESNP model is discussed in Section 4.3. A modification of the ESNP model is described, and the modification is regarded as a shallow model to construct a deep model. Figure 4.6 shows the shallow ESNP model.

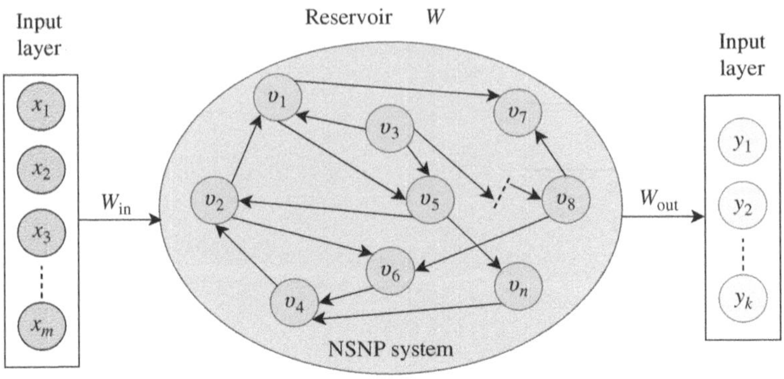

Figure 4.6 A shallow ESNP model [3]/with permission of Elsevier.

The shallow ESNP system has a reservoir of n spiking neurons, the input layer of m neurons, and the output layer of k neurons. In Figure 4.6, a randomly initialized NSNP system is used as the reservoir. Moreover, a linear layer serves as the readout layer and calculates the output of entire network.

For the shallow ESNP system, state and output equations are given by

$$\begin{aligned} \mathbf{v}(\tau) &= \beta_1 \mathbf{v}(\tau-1) + Wf(\beta_2 \mathbf{v}(\tau-1) + W_{\text{in}}\mathbf{x}(\tau)) \\ \mathbf{y}(\tau) &= f_0(W_{\text{out}}\mathbf{v}(\tau)) \end{aligned} \tag{4.33}$$

wherein $\mathbf{v}(\tau) = [v_1(\tau), \ldots, v_n(\tau)]^T$ is the state of neurons at timestep τ; $\mathbf{x}(\tau) = [x_1(\tau), x_2(\tau), \ldots, x_m(\tau)]^T$ is the external input at timestep τ; $\mathbf{y}(\tau) = [y_1(\tau), y_2(\tau), \ldots, y_k(\tau)]^T$ is the system's output at timestep τ; $\beta_1 \in [0, 1]$ and $\beta_2 \in [0, 1]$ are two scalar parameters; $W_{\text{in}} \in R^{m\times n}$ is the input matrix; $W \in R^{n\times n}$ is the interconnect matrix of the reservoir; $W_{\text{out}} \in R^{n\times k}$ is the output matrix; $f(\cdot)$ is a sigmoid function (such as tan$h$), and $f_0(\cdot)$ can be linear or nonlinear functions (for simply, let $f_0(z) = z$).

The output equation is rewritten as

$$\mathbf{y}(\tau) = W_{\text{out}}\mathbf{X}(\tau) \tag{4.34}$$

wherein $\mathbf{X}(\tau) = [\mathbf{x}(\tau)^T, \mathbf{v}(\tau)^T]^T$.

In the execution phase, the initial parameters of the reservoir are randomly configured and the hyperparameter values are set to meet the echo state conditions, and then these parameters are no longer changed. As a result, both the W and W_{in} matrices receive their initial values through random uniform sampling. The learning procedure exclusively updates the weight matrix W_{out} in the final layer. Thus, the ridge regression technique is typically utilized to derive the most suitable output weight matrix W_{out}.

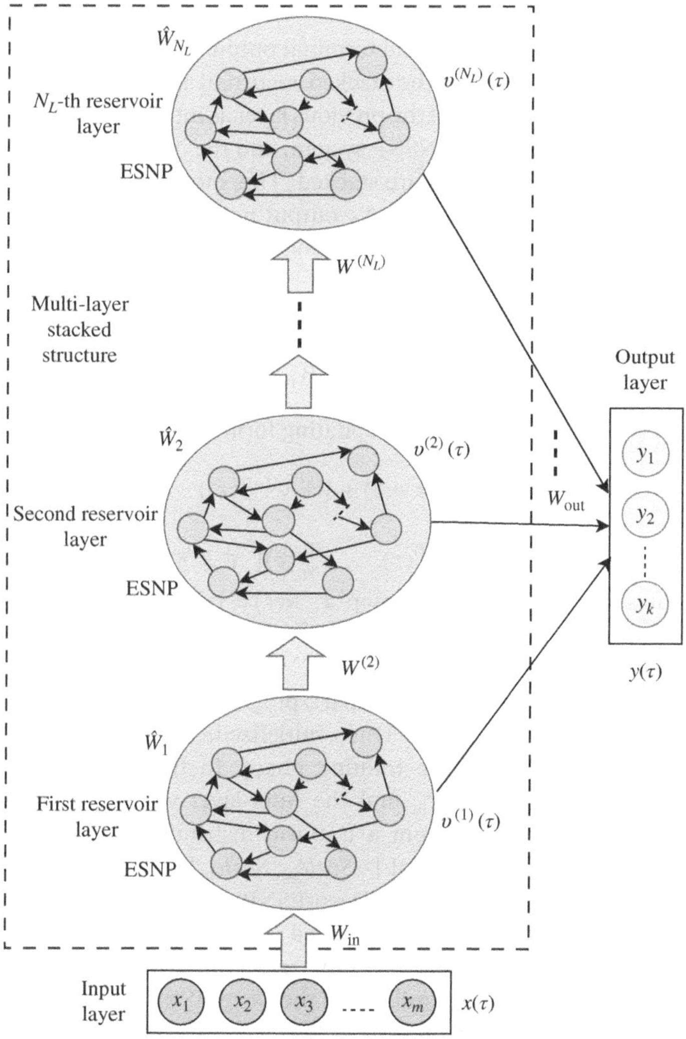

Figure 4.7 The Deep-ESNP model [3]/with permission of Elsevier.

4.5.2 Deep ESNP Model

Drawing from the hierarchical design of deep learning models, a cascaded reservoir configuration with multiple interconnected layers is examined. Figure 4.7 presents the hierarchical Deep-ESNP model, where echo-state computation is achieved through stacked NSNP-based reservoirs [3].

The Deep-ESNP structure inherits from shallow ESNP the fundamental design of having both data-receiving input nodes and terminal output nodes that deliver system results. The reservoirs consist of the stacked recurrent layers, and every recurrent layer contains the output state of the previous layer as input. In the Deep-ESNP model, every recurrent layer is an NSNP system, and the N_L recurrent layers with the same number N_r of neurons are stacked. This model configuration employs N_u input nodes and concludes with N_Y output nodes in its final layer. Consequently, the dynamical equations characterizing the Deep-ESNP architecture are formulated below.

The initial reservoir's state updating formula is expressed by

$$\mathbf{v}^{(1)}(\tau) = \beta_1 \mathbf{v}^{(1)}(\tau - 1)) + \hat{W}_1 f(\beta_2 \mathbf{v}^{(1)}(\tau - 1) + W_{\text{in}} \mathbf{x}(\tau)) \tag{4.35}$$

For other reservoirs ($l = 2, \ldots, N_L$), the state updating formula is given by

$$\mathbf{v}^{(l)}(\tau) = \beta_1 \mathbf{v}^{(l)}(\tau - 1)) + \hat{W}_l f(\beta_2 \mathbf{v}^{(l)}(\tau - 1) + W^{(l)} \mathbf{v}^{(l-1)}(\tau)) \tag{4.36}$$

wherein $l = 2, 3, \ldots, N_L$; $\mathbf{v}^{(l)}(\tau) = [v_1^{(l)}(\tau), \ldots, v_{N_L}^{(l)}(\tau)]^T \in R^{N_R \times 1}$; $\mathbf{v}^{(l)}(\tau)(l = 1, 2, \ldots, N_R)$ are states of neurons at timestep τ; $\mathbf{x}(\tau) = [x_1(\tau), \ldots, x_{N_u}(\tau)]^T \in R^{N_u \times 1}$; $W_{in} \in R^{N_R \times N_v}$, $\beta_1, \beta_2 \in [0, 1]$; $\hat{W}_l \in R^{N_R \times N_R}(l = 1, 2, 3, \ldots, N_L)$; $\hat{W} = [\hat{W}_1, \hat{W}_2, \ldots, \hat{W}_{N_L}] \in R^{N_R \times N_R \times N_L}$, $v^l(\tau) \in R^{N_R \times 1}$; $W^l \in R^{N_R \times N_R}$; $f(\cdot) = \tanh(\cdot)$.

The Deep-ESNP model, like its ESNP counterpart, progresses through sequential initialization and training procedures. Once initialized, the layered reservoir remains unchanged throughout the training phase. In the initialization phase, both the input weight matrix W_{in} and the inter-layer weight matrices $W^{(l)}(l = 2, 3, \ldots, N_L)$ are first sampled from a uniform distribution, then normalized to fall within the bounded interval $[-Scale_{\text{in}}, Scale_{\text{in}}]$. As usual in the standard ESN model, in the evaluation of Deep-ESNP, each reservoir layer considers many different (independently initialized) instances in the hyperparameterization, and all instances have the same reservoir hyperparameters generated with different random seeds. Similarly, in the Deep-ESNP structure, the interconnectedness matrix of each ESNP reservoir, $\hat{W}_l(l = 1, 2, \ldots, N_L)$, undergoes random initialization through distinct uniform distribution seeds.

The case shown in Figure 4.7 is illustrated. In the Deep-ESNP architecture, the first reservoir processes the raw input data fed through the input layer. Once the first reservoir's state is computed at time τ, it serves as the input for the second reservoir's ESNP system, with each subsequent reservoir receiving the prior reservoir's state as its input. Ultimately, the status data from every reservoir is transmitted to the system's final output layer. More precisely, Deep-ESNP's global states emerge from multi-level state compositions, i.e., $V(\tau) = [v^{(1)}(\tau), \ldots, v^{(N_L)}(\tau)]^T \in R^{N_L \times N_R}$. The global state serves as the comprehensive representation of the model, with the time τ output being determined by the following mathematical expression:

$$\mathbf{y}(\tau) = W_{\text{out}} V(\tau) \tag{4.37}$$

where $W_{\text{out}} \in R^{N_Y \times N_L \times N_R}$ is output weight matrix; $V(\tau) \in R^{N_L \times N_R}$ is the global states; $\mathbf{y}(\tau) = [y_1(\tau), \ldots, y_{N_Y}(\tau)]^T \in R^{N_Y}$ is the outputs. For Deep-ESNP, Eq. (4.37) formulates the model's global reservoir state and output weight matrix as a linear combination, representing the aggregated contributions from each reservoir's state.

During the learning process, the output layer of Deep-ESNP undergoes training comparable to that of ESNP, that is, ridge regression algorithm is used to optimize output matrix.

4.5.3 Model Evaluation

4.5.3.1 Time Series Dataset

In the experiment, six time series are adopted as benchmark datasets: MG (Mackey–Glass), Temperature, Sunspot, Brent (Brent oil prices), Lorenz (Lorenz chaos), and Beijing PM2.5.

The dataset on airborne PM2.5 concentrations in Beijing contains comprehensive measurements, featuring eight distinct PM2.5 parameters along with dew point data. This dataset comprises 43,800 entries. The model's training phase utilizes the initial batch of 26,280 data entries. The validation set comprises samples numbered from 26,280 to 35,040, while the test set consists of the remaining 8,760 samples.

The Lorenz chaotic system's data collection contains 3,000 samples, with 2,400 allocated for model training and the remaining 600 reserved for testing purposes.

The Brent dataset comprises 8,216 records of Brent crude oil pricing information. The dataset is partitioned with the initial 4,000 entries allocated for model training, followed by 1,000 entries (4,000–5,000) for validation, while all remaining entries serve as the testing subset.

The Sunspot database compiles 3,132 monthly sunspot counts over 261 consecutive years (1750–2010). The experimental data is partitioned into three subsets: the initial 1,880 data points form the training set, entries from 1,880 to 2,505 serve as the validation set, while all subsequent records are allocated to the test set.

The Temperature dataset included the daily minimum temperatures in Melbourne, Australia, from January 1, 1981 to December 31, 1990 with a total of 3,650 records. In the experiment, the first 1,200 samples are used to train the model; records 1,200–2,400 are the validation set, and the remaining 1,250 records are the test set.

The following equation with time delays generates the Mackey–Glass simulation data [18]:

$$\frac{dx(t)}{dt} = -0.1x(t) + \frac{0.2x(x-\tau)}{1+x^{10}(t-\tau)} \tag{4.38}$$

wherein $17 \leq \tau$. The dynamical system described by Eq. (4.38) demonstrates characteristics of chaos. In the 9,999-point MG dataset, the training subset consists of the first 4,000 points, followed by 1,000 points (4,001–5,000) for validation, while the rest are allocated for testing purposes. The dataset was created with τ set to 17, then scaled to the range between −1 and 1.

For practical applications, a global dataset obtained from the China Meteorological Data Network[1] is employed in meteorological analysis.

For the experimental analysis, meteorological stations throughout Norway provided precipitation records covering the period from early 1998 to mid-2021. After preprocessing and filtering procedures, the final output is designated as the Precipitation dataset. There are a total of 18,355 entries in this dataset. The dataset is partitioned into three segments: the first 11,013 samples form the training set, records numbered 11,014 through 14,684 constitute the validation subset, while all subsequent entries are reserved for testing purposes.

4.5.3.2 Performance Indicators

In the experiment, MSE (mean square error), RMSE (root mean square error) and NRMSE (normalized root mean square error) are adopted to assess the prediction performance of the models, and given as follows:

$$\text{RMSE} = \sqrt{\frac{\sum_{\tau=1}^{T}(x(\tau)-y(\tau))^2}{T}} \tag{4.39}$$

$$\text{MAE} = \frac{\sum_{\tau=1}^{T}|x(\tau)-y(\tau)|}{T} \tag{4.40}$$

$$\text{MSE} = \frac{1}{T}\sum_{\tau=1}^{T}(x(\tau)-y(\tau))^2 \tag{4.41}$$

$$\text{NRMSE} = \frac{\sqrt{\frac{1}{T}\sum_{\tau=1}^{T}(x(\tau)-y(\tau))^2}}{x_{\max}-x_{\min}} \tag{4.42}$$

1 China meteorological data network. https://data.cma.cn/

where $x(\tau)$ indicates the real-world measurement, $y(\tau)$ refers to the estimated value, and T represents the quantity of data points. Typically, prediction models demonstrate improved accuracy and superior performance when their indicator values decrease.

4.5.3.3 Beijing PM2.5 Time Series

In time-series analysis, the Beijing PM2.5 records have become an established benchmark for algorithm validation. During the experimental procedure, every data point within this dataset is normalized to the range of −1 to 1.

Table 4.13 displays the RMSE and MAE results comparing Deep-ESNP's predictive capability against six competing approaches for Beijing's PM2.5 dataset. As shown in Table 4.13, Deep-ESNP achieves an RMSE of 0.0114, outperforming all other compared methods. Among all models evaluated, Deep-ESNP shows the smallest mean absolute error (MAE) at 1.3065e-5, followed by Deep-ESN with a comparatively higher MAE of 8.174e-4. The comparisons demonstrate that the Deep-ESNP outperforms other models in predicting Beijing's PM2.5 levels, as evidenced by its superior RMSE and MAE scores.

4.5.3.4 Lorenz Time Series

The Lorenz dataset is a chaotic dataset, with all the data normalized to [−1, 1] in the experiment.

Performance assessment of the Deep-ESNP approach is conducted using Lorenz system data, with benchmarking against six alternative prediction methods. The performance evaluation of these models is presented in Table 4.14 through RMSE measurements. Comparative analysis reveals that the Deep-ESNP model achieves a substantially smaller RMSE compared to the six benchmark models, demonstrating its superior predictive capability. The results demonstrate that the Deep-ESNP model outperforms others in forecasting the Lorenz chaotic time series.

Table 4.13 Experimental results of the different models on Beijing PM2.5 datasets [3]/with permission of Elsevier.

Model	MAE	RMSE
GBRT	0.1674	0.3156
LSTM	0.1638	0.3310
DA-RNN	0.1592	0.3031
NSNP	0.0461	0.0748
ESNP	0.0147	0.0474
Deep-ESN	8.174e-4	0.3789
Deep-ESNP	1.3065e-5	0.0114

Table 4.14 Experimental results of the different models on Lorenz time series [3]/with permission of Elsevier.

Model	RMSE
nCESN	5.681e-3
BP	1.081e-3
ESN	9.419e-4
mmCESN	2.668e-4
ESNP	1.026e-4
Deep-ESN	3.359e-5
Deep-ESNP	2.640e-6

Table 4.15 Experimental results of the different models on the Brent time series [3]/with permission of Elsevier.

Model	MAE
GBDT	1.38
SVR	1.23
DeepMVE	1.21
ShallowPIPE	1.18
DeepPIPE	1.02
ESNP	0.65
Deep-ESN	0.4326
Deep-ESNP	0.0684

4.5.3.5 Brent Time Series

The experimental analysis utilizes raw, unnormalized Brent crude oil price data in its time series format.

Table 4.15 presents a comparative analysis of MAE results between the Deep-ESNP approach and seven competing models on the Brent dataset. According to Table 4.15, Deep-ESNP achieves a MAE of 0.0684, outperforming all other models in the Brent dataset prediction task and demonstrating superior accuracy. Consequently, among all tested models, Deep-ESN demonstrates superior forecasting performance when applied to the Brent crude oil data.

4.5.3.6 Sunspot Time Series

Sunspot is a classic dataset for prediction tasks, in which all samples are normalized to $[-1, 1]$.

Table 4.16 Experimental results of the different models on Sunspot datasets [3]/with permission of Elsevier.

Model	RMSE	NRMSE
ESNP	0.0104	0.1522
grouped-ESN	—	4.78e-2
MESM	—	1.95e-2
DeeprESN (PCA)	—	1.90e-2
DeeprESN (ELMAE)	—	1.89e-2
Deep-ESN	5.08295e-5	8.4228e-6
Deep-ESNP	2.5408e-6	5.4326e-8

Table 4.16 gives the comparisons of Deep-ESNP model with six methods on the Sunspot time series. The "—" means that the corresponding model does not report the metric value in the original paper. For the RMSE index, Deep-ESNP achieves the lowest forecasting error. In terms of NRMSE, Deep-ESNP demonstrates significantly lower index values compared to alternative models. Consequently, the experimental findings demonstrate that the Deep-ESNP architecture achieves superior forecasting accuracy when applied to sunspot activity data.

4.5.3.7 Temperature Time Series

The experimental study utilizes a weather dataset named Temperature, where all values have been normalized to the range between −1 and 1.

Table 4.17 displays a comparative analysis between the Deep-ESNP approach and seven competing methods using temperature data.

Table 4.17 reveals that among the compared models, Deep-ESNP attains the minimum normalized root mean square error (1.076e-04), with Deep-ESN ranking second at 5.230e-04. Notably, Deep-ESN and Deep-ESNP architectures demonstrate superior predictive performance compared to the remaining six benchmark models. The results demonstrate superior forecasting accuracy of the Deep-ESNP architecture when applied to temperature prediction tasks.

4.5.3.8 MG Time Series

The Mackey–Glass dataset serves as a benchmark standard in chaotic time series forecasting.

The experimental dataset is normalized to a range between −1 and 1.

The performance comparison between Deep-ESNP and five alternative models on the MG dataset is presented in Table 4.18. In Table 4.18, the "—" indicates that the model does not provide an index value in the corresponding original literature. As shown in Table 4.18, the Deep-ESNP model achieves the lowest index values for

Table 4.17 Experimental results of the different models on Temperature time series [3]/with permission of Elsevier.

Model	NRMSE
ESNP	2.9486
ϕ-ESN	1.64e-1
grouped-ESN	1.63e-1
MESM	1.28e-1
DeeprESN (PCA)	1.27e-1
DeeprESN (ELMAE)	1.26e-1
Deep-ESN	5.230e-4
Deep-ESNP	1.076e-4

Table 4.18 Experimental results of the different models on MG dataset [3]/with permission of Elsevier.

Models	NRMSE	RMSE
LSTM-keras	—	0.0056
GRU-keras	—	0.0020
DeeprESN (PCA)	2.79e-2	—
DeeprESN (ELMAE)	2.74e-2	—
Deep-ESN	3.3011e-7	3.7866e-5
Deep-ESNP	1.4998e-7	3.2876e-5

both the RMSE and NRMSE metrics, indicating that it achieves the best prediction performance.

4.5.3.9 Real-life Time Series

The current part utilizes rainfall records sourced from the national weather data platform. Available on the website as a historical collection of global surface weather observations, this dataset contains 18 meteorological variables such as station ID codes, temporal markers (year/month), geographic coordinates (latitude, longitude, elevation), location descriptors (station and country names), along with precipitation records, mean monthly atmospheric pressure, vapor pressure measurements, and temperature averages. Rainfall data spanning January 1998 through May 2021 are collected from multiple weather stations across Norway.

A number of weather observation sites were decommissioned years ago, making their datasets unavailable for analysis, while uncontrollable factors led to the

Table 4.19 Experimental results of the different models on on Precipitation time series [3]/with permission of Elsevier.

Model	NRMSE	MSE	MAE	RMSE
GRU	0.0328	0.0011	0.0160	0.0328
LSTM	1.1571e-5	0.00108	0.0157	0.0329
ESN	1.273	0.00154	0.0293	0.0392
RNN	0.0327	0.00428	0.0311	0.0654
LSTM-SNP	0.0351	0.006327	0.0323	0.0796
GSNP	0.0592	0.014	0.0926	0.1180
ESNP	0.0011	0.000116	0.0007	0.0022
Deep-ESN	5.147e-4	2.49e-10	4.23e-6	9.88e-4
Deep-ESNP	5.53e-6	1.3596e-13	2.26e-8	2.43e-5

permanent loss of several months' worth of measurements. As a result, rainfall data between January 2016 and October 2019 are unavailable in the records. In cases of absent monthly precipitation measurements at specific stations, the dataset employs the average values recorded at all other monitoring sites for the corresponding month as replacements. The processed and filtered data collection is officially named the Precipitation dataset. The preprocessed dataset serves to evaluate how effectively the Deep-ESNP model handles practical applications. The Deep-ESNP model's effectiveness is demonstrated through comparative analysis against eight state-of-the-art recurrent architecture variants Deep-ESN [19], ESNP [1], LSTM-SNP [7], GSNP [8], GRU [20], LSTM [21], RNN [22], and ESN [23].

Table 4.19 gives the comparisons of Deep-ESNP model with the eight prediction methods on the Precipitation time series. For uniformity in analysis, the complete set of experimental data underwent normalization to the $[-1, 1]$ interval.

Four distinct error measurements from the Deep-ESNP model's precipitation predictions are tabulated in the last data row of Table 4.19. The precipitation forecasting results in Table 4.19 reveal the Deep-ESNP's exceptional performance, exhibiting minimal deviations across all four error measures – RMSE (2.43e-5), MAE (2.26e-8), MSE (1.3596e-13), and NRMSE (5.53e-6) – surpassing other comparative models. The precipitation forecasting results in Table 4.19 reveal the Deep-ESNP's exceptional performance, exhibiting minimal values across all four error measures – NRMSE (5.53e-6), MSE (1.3596e-13), MAE (2.26e-8), and RMSE (2.43e-5) – surpassing other comparative models. Based on the evaluation, Deep-ESNP exhibits superior accuracy in predicting precipitation data, confirming its effectiveness in real-world applications.

References

1 L. Long, R. Lugu, X. Xiong, Q. Liu, H. Peng, J. Wang, D. Orellana-Martín, and M.J. Pérez-Jiménez. Echo spiking neural P system. *Knowledge-Based Systems*, 253: 109568, 2022.

2 L. Long, C. Guo, X. Xiong, H. Peng, and J. Wang. A reservoir computing model based on nonlinear spiking neural P systems for time series forecasting. *Applied Soft Computing*, 159:111644, 2024.

3 J. He, H. Peng, J. Wang, and A. Ramírez-de Arellano. A deep echo-like spiking neural P systems for time series prediction. *Knowledge-Based Systems*, 304: 112560, 2024.

4 H. Jaeger and M. Lukoševiěius. Optimization and application of echo state networks with leaky-integrator neurons. *Neural Networks*, 20:335–352, 2007.

5 M. Lukoševĭcius and H. Jaeger. Survey: reservoir computing approaches to recurrent neural network training. *Computer Science Review*, 3(3):127–149, 2009.

6 M.-H. Yusoff, J. Chrol-Cannon, and Y. Jin. Modeling neural plasticity in echo state networks for classification and regression. *Information Sciences*, 364–365:184–196, 2016.

7 Q. Liu, L. Long, Q. Yang, H. Peng, J. Wang, and X. Luo. LSTM-SNP: A long short-term memory model inspired from spiking neural P systems. *Knowledge-Based Systems*, 235:107656, 2022.

8 Q. Liu, L. Long, H. Peng, J. Wang, Q. Yang, X. Song, A. Riscos-Núñez, and M.J. Pérenz-Jiménez. Gated spiking neural P systems for time series forecasting. *IEEE Transactions on Neural Networks and Learning Systems*, 34(9):6227–6236, 2023.

9 S. Yang and J. Liu. Time-series forecasting based on high-order fuzzy cognitive maps and wavelet transform. *IEEE Transactions on Fuzzy Systems*, 26(6): 3391–3402, 2018.

10 S. Bai, J.Z. Kolter, and V. Koltun. An empirical evaluation of generic convolutional and recurrent networks for sequence modeling. *arVix:1803.01271*, 2018.

11 P. Liu, J. Liu, and K. Wu. CNN-FCM: systems modeling promotes stability of deep learning in time series prediction. *Knowledge-Based Systems*, 203:106081, 2020.

12 J.-S. Jang. ANFIS: Adaptive-network-based fuzzy inference system. *IEEE Transactions on Systems, Man, Cybernetics*, 23(3):665–685, 1993.

13 B. Wang, T. Li, Z. Yan, G. Zhang, and J. Lu. DeepPIPE: a distribution-free uncertainty quantification approach for time series forecasting. *Neurocomputing*, 397:11–19, 2020.

14 J. Hu and W. Zheng. Multistage attention network for multivariate time series prediction. *Neurocomputing*, 383:122–137, 2020.

15 Z. Wu, Q. Li, and H. Zhang. Chain-structure echo state network with stochastic optimization: methodology and application. *IEEE Transactions on Neural Networks and Learning Systems*, 33(5):1974–1985, 2021.

16 H. Peng, Z. Lv, B. Li, X. Luo, J. Wang, X. Song, T. Wang, M.J. Pérez-Jiménez, and A. Riscos-Núñez. Nonlinear spiking neural P systems. *International Journal of Neural Systems*, 30(10):2050008, 2020.

17 L. Long, Q. Liu, H. Peng, Q. Yang, X. Luo, J. Wang, and X. Song. A time series forecasting approach based on nonlinear spiking neural systems. *International Journal of Neural Systems*, 32(8):2250020, 2022.

18 M.C. Mackey and L. Glass. Oscillation and chaos in physiological control systems. *Science*, 197:287–289, 1997.

19 C. Gallicchio, A. Micheli, and L. Pedrelli. Design of deep echo state networks. *Neural Networks*, 108:33–47, 2018.

20 J. Chung, C. Gulcehre, K.H. Cho, and Y. Bengio. Empirical evaluation of gated recurrent neural networks on sequence modeling. *Eprint Arxiv*, 2014.

21 S. Hochreiter and J. Schmidhuber. Long short-term memory. *Neural Computation*, 9(8):1735–1780, 1997.

22 J.L. Elman. Finding structure in time. *Cognitive Science*, 14(2):179–211, 1990.

23 H. Jaeger and H. Haas. Harnessing nonlinearity: predicting chaotic systems and saving energy in wireless communication. *Science*, 304(5667):78–80, 2004.

5

Reservoir Computing Models for Time Series Classification

5.1 Introduction

Time series data manifests ubiquitously across diverse real-world domains, including healthcare monitoring [1], financial forecasting [2], manufacturing systems [3], smart mobility solutions [4], as well as activity monitoring encompassing both physical gestures [5], and behavioral patterns [6]. Chronologically ordered observational records constitute what we define as time series. Based on dimensionality, these can be categorized into univariate time series (UTS) and multivariate time series (MTS) sequences. The richer informational content of MTS compared to UTS makes its examination a persistent focus in data science research, with significant applications spanning scientific, technological, and economic domains.

Time series classification (TSC) has been a relatively active research topic in recent decades and has important applications in financial classification [7, 8], traffic classification [9], medical diagnosis [10], and other fields. The classification of multivariate temporal sequences proves valuable for both acoustic signal processing [11] and activity identification [12]. A key limitation lies in effectively encoding the sophisticated time-varying patterns inherent in MTS datasets.

Three principal methodological frameworks dominate current time series classification research: distance-centric, feature-focused, and RC-derived approaches. The academic community has seen several breakthroughs including TDVM (time-expanded SVM with fixed bending distance) [13], MDDTW (Markov-enhanced DTW for multivariate alignment) [14], and DDDTW (derivative-based DTW classifier) [15]. Each contributes unique solutions to temporal pattern analysis challenges through innovative distance metrics and classification architectures.

Several innovative feature-based techniques have emerged in recent research. Ref. [16] developed SMTS, employing symbolic representations for classification purposes. Ref. [17] introduced a novel hybrid model combining LSTM and CNN components with specialized extrusion-excitation modules, termed LSTM-FCN.

Spiking Neural P Systems for Time Series Analysis, First Edition. Jun Wang and Hong Peng.

Other contributions include the COTE framework by Ref. [18] for comprehensive feature detection and CNN-based solutions for MTS classification proposed by [19].

In the context of reservoir computing approaches, the work by Ref. [20] introduced a novel technique utilizing echo state networks combined with the composite adaptive differential evolution (CADE) method. Ref. [21] proposed a classifier that introduced a spatio-temporal aggregation algorithm into the output layer of echo state networks, called FESN. Ref. [22] proposed a reservoir classification method based on MTS representation.

5.2 Preliminaries

5.2.1 Time Series Classification Problems

The objective of time series classification is to predict labels for unknown temporal sequence based on available annotated training instances.

Let $\mathcal{D} = \{(\mathbf{X}_1, \mathbf{y}_1), \ldots, (\mathbf{X}_i, \mathbf{y}_i), \ldots, (\mathbf{X}_T, \mathbf{y}_T)\}$ be a temporal sequence with length T, in which $\mathbf{X}_i = (\mathbf{x}_i(1), \ldots, \mathbf{x}_i(\tau), \ldots, \mathbf{x}_i(\tau_{\max}))^T$ is the ith data, and $\mathbf{x}_i(\tau) \in \mathbb{R}^D$, $\tau = 1, 2, \ldots, \tau_{\max}$; $\mathbf{y}_i \in R^{N_c}$ is the corresponding label; D is the dimension number of data; $\tau_{\max}$ is the length of time series, and N_c denotes number of categories.

Most TSC approaches adopt an encoder–decoder paradigm where the encoder learns latent features from sequential data, and the decoder performs probabilistic classification using these extracted features. The encoder captures temporal dependencies, while the decoder maps these dependencies to output distributions. RNNs excel at processing sequential patterns, making them ideal encoder candidates for TSC applications. For a multivariate time series input $\mathbf{X} = (\mathbf{x}(1), \ldots, \mathbf{x}(\tau), \ldots, \mathbf{x}(\tau_{\max}))^T$, the dynamical state transition follows the equation:

$$\mathbf{v}(\tau) = f_{\text{en}}(\mathbf{x}(\tau), \mathbf{v}(\tau - 1); \theta_{\text{en}}) \tag{5.1}$$

where $\mathbf{v}(\tau)$ is the state of the model, and θ_{en} is the encoder's parameter.

Let $\mathbf{V} = (\mathbf{v}(1), \mathbf{v}(2), \ldots, \mathbf{v}(\tau_{\max}))^T$ be a state sequence of the RNN at time $1, 2, \ldots, \tau_{\max}$. $\mathbf{V}$ can be used to extract the representation $\mathbf{r}_\mathbf{X} = r(\mathbf{V})$ of the input $\mathbf{X}$. To reconstruct the original input, an RNN often uses the last state, that is, $\mathbf{r}_\mathbf{X} = \mathbf{v}(\tau_{\max})$.

The decoder is used to map the representation $\mathbf{r}_\mathbf{X}$ to the output space, which is a set of class labels $\mathbf{y}$ for TSC task,

$$\mathbf{y} = f_{\text{de}}(\mathbf{r}_\mathbf{X}; \theta_{\text{de}}) \tag{5.2}$$

where $f_{\text{de}}(\cdot)$ can be a feedforward neural network or linear model, and θ_{de} denotes the decoder's parameter.

In the context of a completely trainable recurrent neural network, when provided with multivariate time series data, $\mathbf{X}_1, \mathbf{X}_2, \ldots, \mathbf{X}_T$, and the corresponding labels $\mathbf{y}_1, \mathbf{y}_2, \ldots, \mathbf{y}_T$, the encoder parameters θ_{en} and decoder parameters θ_{de} can be learned through minimization of the specified empirical loss function:

$$\theta_{\text{en}}^*, \theta_{\text{de}}^* = \arg\min_{\theta_{\text{en}}, \theta_{\text{de}}} \frac{1}{T} \sum_{\tau=1}^{T} l(\mathbf{y}_\tau, f_{\text{de}}(r(f_{\text{en}}(\mathbf{X}_\tau, \mathbf{V})))) \tag{5.3}$$

where the function $l(\cdot, \cdot)$ serves as a general-purpose loss metric, particularly applicable as label-wise cross entropy in TSC tasks. Consequently, the BPTT method enables identification of the optimal solution, θ_{en}^* and θ_{de}^*.

In TSC tasks, standard recurrent neural networks (RNNs) face challenges when dealing with extended sequences (denoted by large T values), primarily due to issues like vanishing gradients in deep architectures. To address these limitations, advanced RNN variants including long short-term memory (LSTM) [23] and gated recurrent units (GRU) [24] have been introduced.

5.2.2 Reservoir Computing and Model Space Representation

As an innovative recurrent neural network architecture [25], reservoir computing (RC) employs distinct design principles compared to conventional RNNs. Its framework consists of two key components: (i) a dynamic reservoir that nonlinearly transforms input signals, and (ii) a trainable readout mechanism typically implemented as either a linear model or shallow neural network. Notably, RC's unique training paradigm fixes the reservoir's randomly initialized weights while exclusively optimizing the readout parameters.

As a subclass of reservoir computing (RC) architectures, echo state networks (ESNs) are particularly suited for regression tasks, with variants such as the leaky integrator model documented in Ref. [26]. The dynamical system's state transition in the reservoir layer follows the equation:

$$\mathbf{v}(\tau) = (1 - \lambda)\mathbf{v}(\tau - 1) + \lambda \tanh(\mathbf{W}_{\text{in}}\mathbf{x}(\tau) + \mathbf{W}\mathbf{v}(\tau - 1) + \mathbf{b}) \tag{5.4}$$

In the encoding module, the learnable parameters include: the input transformation matrix $\mathbf{W}_{\text{in}}$, the recurrent connection matrix $\mathbf{W}$ within the reservoir, the bias term $\mathbf{b}$, along with a leakage factor λ whose value ranges from zero to unity. Encoder weights are stochastically initialized and kept fixed throughout the processing phase. The ESN framework formulates the decoder through

$$\mathbf{y} = f_{\text{de}}(\mathbf{r}_{\mathbf{X}}) = \mathbf{W}_{\text{out}}\mathbf{r}_{\mathbf{X}} + \theta_{\text{out}} \tag{5.5}$$

The parameters of the decoder $\theta_{de} = \{\mathbf{W}_{out}, \theta_{out}\}$ can be learnt by minimizing the following loss function

$$\mathbf{W}^*_{out}, \theta^*_{out} = \arg \min_{\mathbf{W}_{out}, \theta_{out}} \frac{1}{2}||(\mathbf{W}_{out}\mathbf{r}_\mathbf{X} + \theta_{out}) - \mathbf{y}||^2 + \lambda||\mathbf{W}_{out}||^2 \tag{5.6}$$

Within the echo state network framework, the feature vector $\mathbf{r}_\mathbf{X}$ corresponds to the model's output space, specifically expressed as $\mathbf{r}_\mathbf{X} = \mathbf{v}(\tau)$. Consequently, the decoding mechanism can be characterized by

$$\mathbf{y}(\tau) = \mathbf{W}_{out}\mathbf{r}_\mathbf{X} + \theta_{out} = \mathbf{W}_{out}\mathbf{v}(\tau) + \theta_{out} \tag{5.7}$$

5.3 Basic Models

5.3.1 NSNP Systems

NSNP systems [27] represent an enhanced version of standard SNP systems where nonlinear mechanisms govern the spiking behavior. In Chapter 2, NSNP systems are introduced as a theoretical computing model. When addressing time series classification challenges, the foundational computational framework often requires expansion along with certain restriction adjustments. This adapted version continues to be termed NSNP systems for practical purposes.

NSNP systems possess a unique nonlinear spiking characteristic absent in conventional SNP models. This mechanism operates through dual components: consumption and generation of spikes with nonlinear properties. Governed by this framework, NSNP exhibits complex nonlinear behavioral patterns.

Definition 5.1 A NSNP system is a tuple

$$\Pi = (O, \sigma_1, \sigma_2, \ldots, \sigma_n, syn, \mathbf{x})$$

wherein

1. $O = \{\alpha\}$ is a singleton alphabet, where α is called the spike;
2. $\sigma_1, \sigma_2, \ldots, \sigma_n$ are n neurons, with $\sigma_i = (v_i, r_i), i = 1, 2, \ldots, n$, wherein
   a. $v_i \in \mathbb{R}$ is the initial state of neuron σ_i;
   b. r_i is nonlinear spiking rule, $\alpha^{g(v)} \to \alpha^{f(v)}$, in which functions $g(\cdot)$ and $f(\cdot)$ are nonlinear; $v = x_i + v_i$, in which x_i denotes ith external input.
3. $syn \subseteq \{1, 2, \ldots, n\} \times \{1, 2, \ldots, n\}$ (synapses).
4. $\mathbf{x} = (x_1, x_2, \ldots, x_D)$ indicates the external inputs.

Illustrated in Figure 5.1, the NSNP system denoted by Π contains a total of n neurons, $\sigma_1, \sigma_2, \ldots, \sigma_n$. Topologically speaking, the NSNP framework constitutes a directed network exhibiting comprehensive structural connectivity. This system

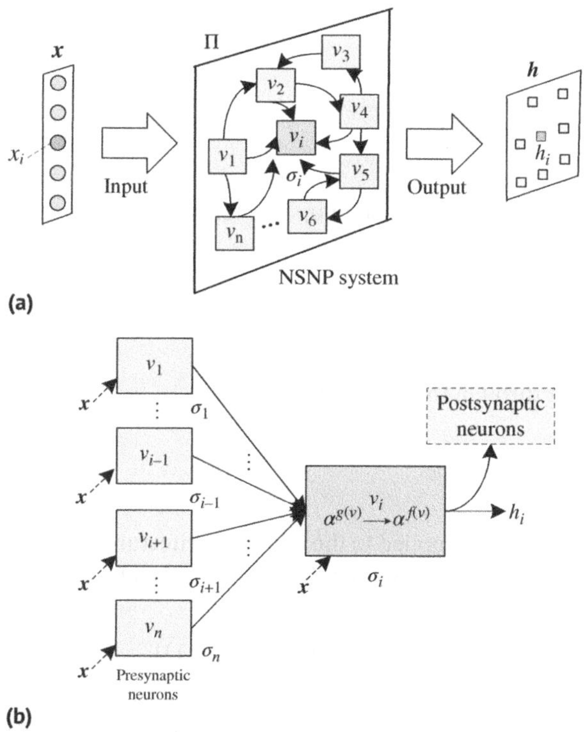

Figure 5.1 (a) NSNP system Π; (b) Neuron σ_i and its presynaptic neurons [28]/with permission of Elsevier.

operates as a MIMO (multi-input multi-output) configuration where all neural units simultaneously serve as both input and output nodes. The NSNP architecture incorporates neurons where each σ_i possesses both a state parameter v_i and a nonlinear spike generation rule $\alpha^{g(v)} \to \alpha^{f(v)}$. Suppose that $v_i(\tau)$ is the state of neuron σ_i at time τ. The semantics of the spiking rule can be described as follows. When neuron σ_i fires, the spikes with value $g(v)$ are consumed (and spikes with value $v_i(\tau) - g(v)$ remain), and then the spikes with value $f(v)$ are generated and sent to its postsynaptic neurons and regarded as its external output $h_i(\tau)$, where $v = \hat{x}(\tau) + v_i(\tau - 1)$, $v_i(\tau - 1)$ is its state at last time $\tau - 1$, and $\hat{x}(\tau)$ is the current external input. Therefore, neuron σ_i's state update and output formulations are presented as

$$v_i(\tau) = v_i(\tau - 1) - g\left(\hat{x}(\tau) + v_i(\tau - 1)\right) + \sum_{j=1, i\neq j}^{n} f\left(\hat{x}(\tau) + v_j(\tau - 1)\right) \tag{5.8}$$

$$h_i(\tau) = f\left(\hat{x}(\tau) + v_i(\tau - 1)\right) \tag{5.9}$$

where $i = 1, 2, \ldots, n$. To simplify the complexity of model construction, let $g(v) = -f(v)$. As a result, we obtain

$$v_i(\tau) = v_i(\tau - 1) + \sum_{j=1}^{n} f(\hat{x}(\tau) + v_j(\tau - 1)) \tag{5.10}$$

Treating $\hat{x} = \sum_{s=1}^{D} x_s$ as an external input, the state update formula for neuron σ_i can be reformulated as:

$$v_i(\tau) = v_i(\tau - 1) + \sum_{j=1}^{n} f(\sum_{s=1}^{D} x_s(\tau) + v_j(\tau - 1)) \tag{5.11}$$

Neuron σ_i computes its output through the following equation:

$$h_i(\tau) = f(\sum_{s=1}^{D} x_s(\tau) + v_i(\tau - 1)) \tag{5.12}$$

With further parameter adjustments applied to the NSNP architecture, the neuronal state dynamics of σ_i can be expressed in the following modified form:

$$v_i(\tau) = \beta_1 \cdot v_i(\tau - 1) + \sum_{j=1}^{n} w_{ij}^1 \cdot f(\sum_{s=1}^{D} w_{is}^2 \cdot x_s(\tau) + \beta_2 \cdot v_j(\tau - 1)) \tag{5.13}$$

where $\beta_1, \beta_2, w_{ij}^1$, and w_{is}^2 are the parameters. The mathematical representation for the output of σ_i can be restructured by

$$h_i(\tau) = f(\sum_{s=1}^{D} w_{is}^3 \cdot x_s(\tau) + \gamma \cdot v_i(\tau - 1)) \tag{5.14}$$

where w_{is}^3 and γ are the parameters, and $i = 1, 2, \ldots, n$.

Let $\mathbf{v}(\tau) = [v_1(\tau), v_2(\tau), \ldots, v_n(\tau)]^T$, $\mathbf{x}(\tau) = [(x_1(\tau), x_2(\tau), \ldots, x_D(\tau)]^T$, and $\mathbf{h}(\tau) = [h_1(\tau), h_2(\tau), \ldots, h_m(\tau)]^T$. As a result, the dynamic state update of the system is described by

$$\mathbf{v}(\tau) = \beta_1 \cdot \mathbf{v}(\tau - 1) + Wf(W_{\text{in}}\mathbf{x}(\tau) + \beta_2 \cdot \mathbf{v}(\tau - 1)) \tag{5.15}$$

where $\beta_1, \beta_2 \in (0, 1]$ are scalar parameters, and $W = (w_{ij}^1)_{m\times m}$ and $W_{\text{in}} = (w_{ij}^2)_{n\times m}$ are parameter matrices. The generated result from the system is formulatable as

$$\mathbf{h}(\tau) = f(W'_{\text{out}}\mathbf{x}(\tau) + \gamma \cdot \mathbf{v}(\tau - 1)) \tag{5.16}$$

where $W'_{\text{out}} = (w_{ij}^3)_{n\times m}$ is the matrix, and $\gamma \in (0, 1]$ is the scalar factor; set $f(\cdot) = \tanh(\cdot)$.

Following the processes of specialization and parameterization, the resulting architectures are designated as extended NSNP systems. These modified systems,

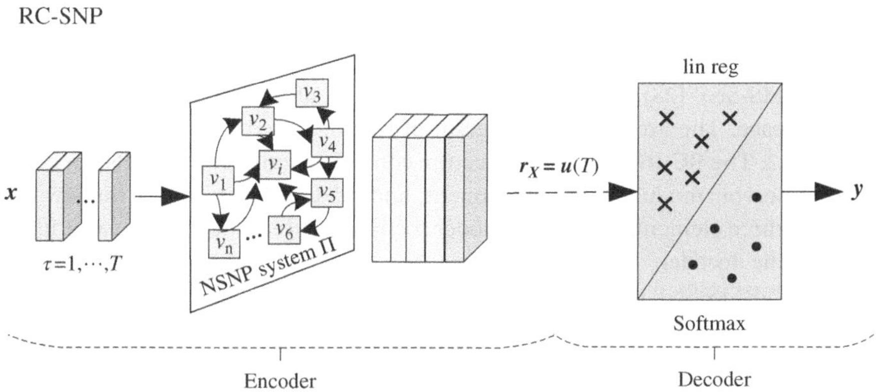

Figure 5.2 RC-SNP model [28]/with permission of Elsevier.

differing from the initial NSNP models [27], incorporate both simplifications and extensions. The extended variant features homogeneous neurons where each unit possesses identical state components and adheres to uniform state update dynamics. The mathematical foundations of these systems are established through Eqs. (5.15) and (5.16).

5.3.2 RC-SNP Model

Building upon the enhanced NSNP framework, a novel RC architecture tailored for TSC applications is discussed, termed the RC-SNP approach [28]. The structural composition of this model is illustrated in Figure 5.2. The RC-SNP model is composed of an input layer, reservoir, output model space, and readout layer. The input layer, reservoir, and output model space constitute the encoder, and the decoder is the readout layer.

The RC-SNP model distinguishes itself from conventional RC approaches by employing an NSNP system as its reservoir. The inherent nonlinear spiking characteristics of NSNP systems enable complex dynamic behaviors, making them particularly suitable for reservoir computing applications. In NSNP reservoir systems, the state update dynamics follow Eq. (5.15). The encoder's weights θ_{en}, comprising $\mathbf{W}_{in}$ and $\mathbf{W}$, are assigned random initial values and remain static during operation.

The reservoir system's state trajectory, denoted as $\mathbf{V} = (\mathbf{v}(1), \mathbf{v}(2), \ldots, \mathbf{v}(\tau_{max}))^T$, demonstrates complex dynamical behaviors. Within the RC-SNP framework, the final state vector $\mathbf{v}(\tau_{max}))^T$ serves as the input's feature representation, expressed as $\mathbf{r}_X = \mathbf{v}(\tau_{max})$.

In the RC-SNP framework, the decoding process employs linear regression, while MTS categorization is achieved through softmax activation. The parameters of the decoder $\theta_{de} = \{\mathbf{W}_{out}, \theta_{out}\}$ can be trained by a ridge regression algorithm.

5.3.3 RC-RMS-SNP Model

This section introduces an enhanced version of the RC-SNP model, designated as RC-RMS-SNP [28], which incorporates RMS to substitute the original output model space. The architectural diagram of this modified model is displayed in Figure 5.3. The RC-RMS-SNP architecture comprises four key elements: an input layer, reservoir module, RMS component, and readout layer. In this framework, the first three elements form the encoder section, whereas the readout layer functions as the decoder.

The RC-RMS-SNP model incorporates RMS, distinguishing it from the RC-SNP variant. Below is an explanation of RMS.

Reservoir computing systems derive their generative capacity from an extensive repertoire of input-driven dynamical states. For forecasting subsequent inputs, specific dynamical modes are selectively activated based on the required prediction time scale. Therefore, the RMS has been introduced to generate the presentation of the input $\mathbf{r}_X$ for MTS [22]. In the RMS, each MTS is expressed by the parameters of a linear model, which predicts the next reservoir state by considering all the reservoir dynamics.

For extracting the latent representation $\mathbf{r}_X$, each multivariate time series undergoes processing through the reservoir/NSNP architecture. The forward prediction mechanism for subsequent reservoir states is then defined by

$$\mathbf{v}(\tau + 1) = V_\nu \mathbf{v}(\tau) + v_0 \tag{5.17}$$

By applying ridge regression's objective function minimization, one can obtain solutions for parameters V_ν and v_0:

$$\theta_0^* = \arg\min_{\mathbf{V}_\nu, v_0} \frac{1}{2} ||(\mathbf{V}_\nu \mathbf{v}(\tau) + v_0 - \mathbf{v}(\tau + 1)||^2 + \lambda ||\mathbf{V}_\nu||^2 \tag{5.18}$$

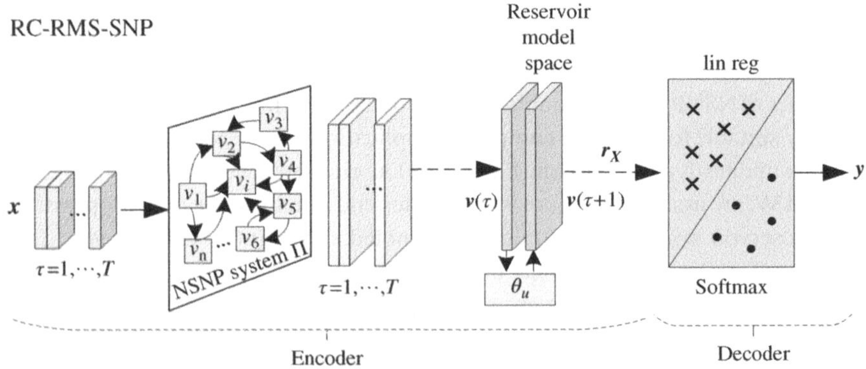

Figure 5.3 RC-RMS-SNP model [28]/with permission of Elsevier.

The MTS is mathematically expressed as $\mathbf{r_X} = \theta_v = [vec(V_v); v_0]$, known as the RMS formulation. This formulation effectively models the generative process of reservoir state sequences, $p(\mathbf{v}(\tau_{\max}), \mathbf{v}(\tau_{\max} - 1)), \ldots, \mathbf{v}(1); \mathbf{r_X})$.

The RC-RMS-SNP approach, building upon the RC-SNP framework, employs linear regression for modeling while utilizing softmax for MTS categorization. Consequently, decoder parameters are optimized through standard ridge regression techniques.

Both RC models are developed in Python utilizing the Scikit-learn machine learning framework, with model parameters being optimized through ridge regression.

Both reservoir computing architectures employ NSNP systems as their dynamical reservoir core. Consistent with conventional RC design principles, critical parameters like input scaling, connection density, and spectral radius factors are carefully tuned to guarantee the echo state property.

5.4 Improved Models

5.4.1 Representation of Reservoir Model Space

Within reservoir computing frameworks, the reservoir model space (RMS) approach [22] operates without supervision to encode temporal sequences into vector representations through linear model parameterization. Herein, RMS is employed to quantify neural state representations in the NSNP architecture.

Within RC architectures, the final readout layer (termed the decoder for classification purposes) frequently utilizes a linear structure:

$$y = q(V_r) = H_x V_r + h_x \tag{5.19}$$

The parameters in readout layer, $\delta_{dec}(H_x, h_x)$, can be learned by minimizing ridge regression loss function:

$$\delta^*_{dec} = \arg \min_{\{H_x, h_x\}} \frac{1}{2} ||(H_x V_r + h_x) - y||^2 + \lambda ||H_x||^2 \tag{5.20}$$

Within the reservoir modeling framework, every temporal sequence is characterized through linear model coefficients that forecast subsequent reservoir conditions based on present system states:

$$\mathbf{v}(\tau) = H_v \mathbf{v}(\tau) + h_v \tag{5.21}$$

The formation procedure of RMS representation is illustrated in Figure 5.4. Consequently, the reservoir's temporal state can be mathematically defined as $V_r = \theta_v = [vec(H_v); h_v] \in \mathbb{R}^{R(R+D)}$ for RMS. During processing, the input data is

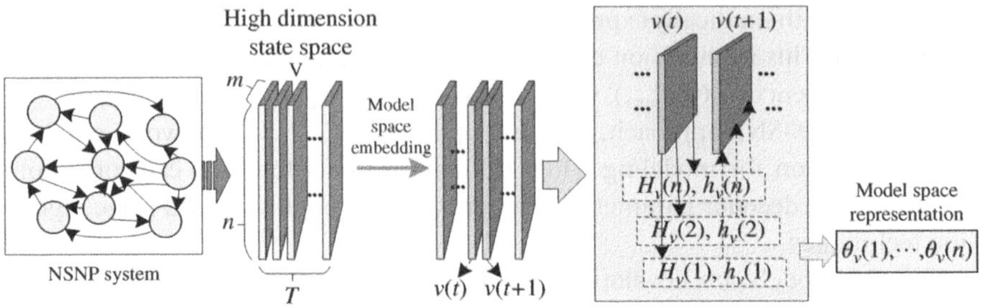

Figure 5.4 Reservoir model space for NSNP system [29]/with permission of Elsevier.

structured as a 3D tensor $U \in \mathbb{R}^{m \times T \times n}$, with dimensions representing: m for training/test samples, T for the maximal temporal sequence length, and n indicating reservoir computing units. Figure 5.4 shows that n indicates the coronal plane of the RMS, with T marking its axial plane. The sequence of parameters from $\theta_v[1]$ to $\theta_v[n]$ thereby constitutes the input's RMS-based characterization.

5.4.2 Reduction Dimensionality

Within reservoir computing architectures, input signals undergo nonlinear projection into higher-dimensional states, after which a computationally efficient linear readout mechanism processes these states for final decision-making. The reservoir's property of generating high-dimensional state sequences may prolong the training duration of the readout layer's linear model. Moreover, the significant increase in the number of parameters in a linear model can increase the difficulty of RMS representation during processing.

In order to reduce the dimensionality of the reservoir, dimensionality reduction methods can be adopted to map data from high-dimensional space to low-dimensional space. Unsupervised techniques for reducing dimensionality, including PCA and its kernel-based variant, offer viable solutions. PCA operates by mapping data points along the dominant eigenvector derived from covariance analysis. When implemented on reservoir computing (RC) systems, this transformation effectively compresses the state space while enhancing computational performance. Consequently, RC frameworks frequently employ PCA to optimize their state representations.

5.4.3 TSC-DR-NSNP Model

Within the TSC-NSNP framework, the NSNP reservoir performs dimensional expansion, converting low-dimensional input sequences into enriched high-dimensional representations through state space transformation. Then, a linear

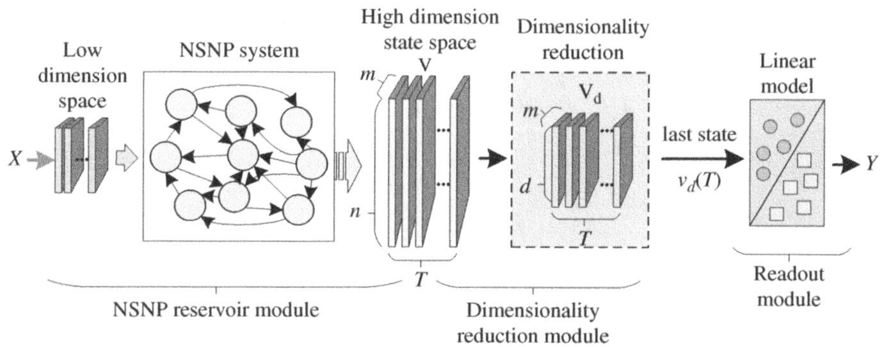

Figure 5.5 The structure of TSC-DR-NSNP model [29]/with permission of Elsevier.

model is used for training and prediction in the output layer. High dimensional data increases the training and prediction time of the model, and the processing of high dimensional data is more complex than that of low dimensional data.

To streamline the ridge regression algorithm's data processing and accelerate both training and prediction phases, PCA is integrated into the TSC-NSNP framework between the NSNP reservoir and output layer for state space dimension reduction. The dimensionality-reduced variant of TSC-NSNP is designated as TSC-DR-NSNP [29], shown in Figure 5.5.

The NSNP reservoir generates a temporal state sequence represented as a 3D array $V \in \mathbb{R}^{n \times T \times m}$. Following PCA, the original matrix V is projected into a lower-dimensional space, resulting in $V_d \in \mathbb{R}^{n \times T \times d}$, where d denotes the reduced subspace dimension satisfying $d < n$. Since PCA dimensionality reduction exclusively operates on 2D data structures, preprocessing of U becomes essential. The initial step involves reshaping V into a 2D matrix $V' \in \mathbb{R}^{NT \times R}$. PCA is subsequently applied to matrix V', yielding the reduced-dimensional tensor $V'' \in \mathbb{R}^{NT \times D}$. Ultimately, the reduced 2D tensor V'' is reshaped into its 3D counterpart V_d.

As illustrated in Figure 5.5, the TSC-DR-NSNP approach employs ridge regression on the final state of the compressed reservoir space V_d for model training and subsequent classification prediction.

5.4.4 TSC-RMS-NSNP Model

The ridge regression algorithm in TSC-DR-NSNP's output layer is trained using the final temporal state generated through NSNP reservoir computation and subsequent dimension compression. Note that features that are not particularly important for prediction tasks can still be used to characterize time series. The TSC-DR-NSNP architecture incorporates a reservoir model space (RMS) bridging the dimensionality reduction and readout components. This intermediate layer

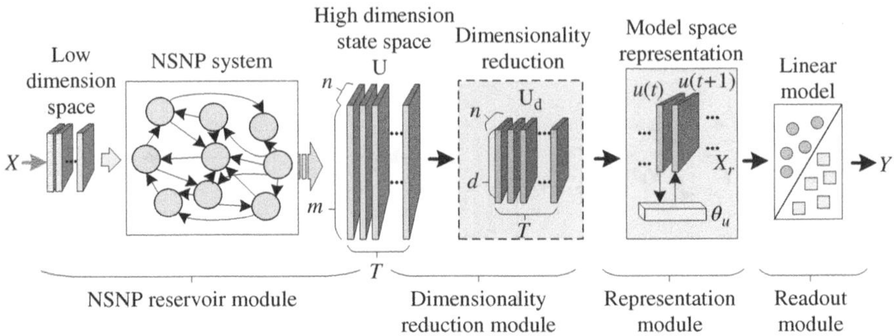

Figure 5.6 The structure of TSC-RMS-NSNP model [29]/with permission of Elsevier.

transforms the compressed NSNP reservoir states into an alternative form called X_r, enabling full exploitation of temporal sequence information. When reservoir model space is integrated into TSC-DR-NSNP, the resulting framework is known as TSC-RMS-NSNP [29], shown in Figure 5.6.

Figure 5.6 demonstrates how the NSNP reservoir transforms low-dimensional time series into high-dimensional representations, ultimately producing the reduced-dimensional state of the complete reservoir system. Unlike conventional linear models relying on the last observed state, the TSC-RMS-NSNP framework maintains the complete NSNP reservoir state and employs ridge regression to forecast subsequent states from current data.

5.4.5 The Implementation of RC Models

The three described models are the variants of RC models, including TSC-NSNP, TSC-DR-NSNP, and TSC-RMS-NSNP models. Therefore, the three RC models can be implemented in the RC framework. TSC-NSNP is a base model. Building upon the TSC-NSNP framework, the TSC-DR-NSNP variant incorporates dimensionality reduction methods. Subsequently, the TSC-RMS-NSNP architecture further enhances this approach by integrating RMS expression capabilities.

Similar to conventional RC frameworks, these models follow a three-stage implementation process:

1. Initialization stage:
   At the initialization step, the weights for inputs and reservoir nodes receive stochastic initial values that stay fixed throughout subsequent operations.
2. RMS construction stage:
   The TSC-RMS-NSNP approach employs ridge regression to generate RMS representations of time series data, while both TSC-NSNP and TSC-DR-NSNP methodologies omit this computational step.

3. Training stage:
   The output layer's parameters are optimized through ridge regression, while classification is achieved by employing the softmax function.

5.5 Model Evaluation

Four different RC architectures are tested in the experiments: RC-SNP, RC-RMS-SNP, TSC-RMS-NSNP, and TSC-DR-NSNP. The four reservoir computing architectures undergo performance assessment across 17 standard time series datasets, with their classification capabilities benchmarked against 14 state-of-the-art or baseline algorithms.

5.5.1 Datasets

Seventeen publicly available reference datasets, sourced from the UCI,[1] UCR, and UEA,[2] serve as evaluation benchmarks for time series categorization tasks. There are three UTS: Swedish Leaf, Chlorine Concentration (Chlo Conc), and Distal Phalanx Outline (Dist Phal). The other 14 datasets are MTS: UWave Gesture Library (uWave), Libras, Electrocardiography (ECG), Robotic Arm Failure (RobotFail), CMUsubject16 (CMU), Walk versus Run (WALK), Kick versus Punch (KICK), Performance Measurements (PEMS), Japanese Vowels (Jp. Vow.), Wafer, Character Trajectory (Ch. Traj.), Network Flow (NetFlow), Arabic Digits (Arab. Dig.), and Australian Sign Language (Auslan).

Dataset specifications can be found in Table 5.1. During experimental procedures, every time series undergoes transformation into a 3D tensor structure before being fed into the model for both learning and forecasting purposes.

5.5.2 Performance Metric

The four RC models are assessed against current MTS classifiers using accuracy rate as the primary performance metric, where accuracy is mathematically expressed as

$$\text{Accurate rate} = \frac{\text{Number of correctly classified samples}}{\text{Total number of test samples}} \tag{5.22}$$

1 https://archive.ics.uci.edu/datasets
2 https://www.timeseriesclassification.com/dataset.php

Table 5.1 Statistical information of 17 time series datasets [29]/with permission of Elsevier.

Dataset	#Variables	#Classes	Length	Test	Train	Source
Arab. Dig.	13	10	[4,93]	2,200	6,600	UCI
Auslan	22	95	[45,136]	1,425	1,140	UCI
NetFlow	4	13	[50,994]	534	803	UEA
Swedish Leaf	1	15	[128,128]	625	500	UCR
Chlo Conc	1	3	[166,166]	3,840	467	UCR
Dist Phal	1	3	[80,80]	139	400	UCR
Ch.Traj.	3	20	[109,205]	2,558	300	UCI
Wafer	6	2	[104,198]	896	298	UCR
Jp. Vow.	12	9	[7,29]	370	270	UCI
PEMS	963	7	[144,144]	173	267	UCI
uWave	3	8	[315,315]	427	200	UCR
Libras	2	15	[45,45]	180	180	UCI
ECG	2	2	[39,152]	100	100	UCR
RobotFail	6	4	[15,15]	64	100	UCI
CMU	62	2	[127,580]	29	29	UEA
WALK	62	2	[128,1918]	16	28	UEA
KICK	62	2	[274,841]	10	16	UEA

5.5.3 Comparison Results

A comparative analysis is conducted between four RC architectures and six RNN-inspired classifiers using 17 benchmark datasets for time series classification. The 6 RNN-inspired classifiers are dr-rmESN [22], bi-rmESN [22], rmESN [22], DWT-1NN [14], GRU [24], and LSTM [23]. Table 5.2 displays the accuracy performance of four RC architectures against six RNN-style sequential classifiers. The results of the 6 time series classifiers are retrieved from their original references, and their hyper-parameters can be found in the original references.

From Table 5.2, it can be observed that these RC models achieve the better classification accuracy on most datasets. The RC-RMS-SNP model is an improvement of the RC-SNP model with the embedding of the RMS. Table 5.2 reveals that: (i) The RC-RMS-SNP model outperforms others on nine benchmark datasets (Arab. Dig, Auslan, Ch. Traj., Wafer, Jp. Vow., uWave, Libras, ECG, and WALK); (ii) For the CMU dataset, dr-rmESN(Max) shows superior performance with a 7.24% advantage over RC-RMS-SNP; (iii) On PEMS dataset, GRU achieves 17.36% higher accuracy than RC-RMS-SNP; (iv) For the remaining five datasets, RC-RMS-SNP's performance gaps with top models are all within 6%.

Table 5.2 The accuracy rate values of the different models [29]/with permission of Elsevier.

Dataset	DWT-1NN	GRU	LSTM	bi-rmESN	rmESN
Swedish Leaf	81.72	88.24 ± 01.62	89.58 ± 00.71	96.25 ± 0.66	94.51 ± 00.80
Chlo Conc	62.6	60.85 ± 1.13	60.42 ± 1.10	83.72 ± 0.62	85.60 ± 0.41
Dist Phal	74.82	72.66 ± 1.29	70.94 ± 2.93	75.23 ± 0.73	75.57 ± 1.32
Ch.Traj.	95.78	70.79 ± 17.71	37.10 ± 14.62	97.00 ± 0.11	97.36 ± 0.24
Wafer	95.09	98.41 ± 0.86	96.32 ± 3.70	97.01 ± 0.40	97.78 ± 0.29
Jp. Vow.	93.51	94.00 ± 2.21	92.70 ± 1.36	97.62 ± 0.46	97.83 ± 0.50
PEMS	70.52	89.67 ± 1.51	85.57 ± 1.57	72.37 ± 2.02	70.40 ± 3.79
uWave	89.46	79.49 ± 2.65	72.52 ± 1.71	90.51 ± 1.16	88.88 ± 0.52
Libras	88.33	71.56 ± 4.60	68.22 ± 2.62	86.00 ± 0.65	88.11 ± 1.43
ECG	84	81.20 ± 3.49	76.20 ± 4.26	85.20 ± 0.40	85.20 ± 0.75
RobotFail	68.75	63.75 ± 2.30	64.69 ± 3.22	56.88 ± 1.25	64.38 ± 1.17
CMU	100	54.79 ± 9.52	55.17 ± 13.79	94.83 ± 1.72	98.27 ± 1.72
WALK	100	75.57 ± 0.13	75.57 ± 0.13	100.0 ± 0.00	100.0 ± 0.00
KICK	70	35.83 ± 5.97	45.67 ± 5.46	100.0 ± 0.00	75.02 ± 4.83
NetFlow	92.08	93.13 ± 2.25	91.84 ± 1.39	96.75 ± 0.50	96.96 ± 0.54
Auslan	85.61	1.05 ± 0.00	1.05 ± 0.00	97.64 ± 0.35	97.25 ± 0.25
Arab. Dig.	—	95.98 ± 2.91	96.61 ± 0.69	96.46 ± 0.44	98.12 ± 0.21

Dataset	dr-rmESM	RC-RMS-SNP	RC-SNP	TSC-DR-NSNP	TSC-RMS-NSNP
Swedish Leaf	95.56 ± 0.66	93.79 ± 00.47	83.54 ± 01.17	80.45 ± 1.44	93.87 ± 0.47
Chlo Conc	85.07 ± 1.36	80.56 ± 0.57	60.71 ± 0.84	58.26 ± 0.15	74.59 ± 0.79
Dist Phal	76.52 ± 0.70	76.04 ± 0.65	74.60 ± 0.56	73.96 ± 1.11	76.91 ± 1.56
Ch.Traj.	97.52 ± 0.54	98.48 ± 0.21	93.03 ± 0.34	93.62 ± 0.68	98.55 ± 0.11
Wafer	96.85 ± 0.65	99.20 ± 0.14	98.46 ± 0.24	98.50 ± 0.25	99.27 ± 0.12
Jp. Vow.	97.99 ± 0.65	98.14 ± 0.35	97.30 ± 0.40	96.68 ± 0.44	97.95 ± 0.53
PEMS	69.02 ± 3.48	72.31 ± 2.56	76.24 ± 2.94	72.77 ± 3.44	73.93 ± 1.72
uWave	89.39 ± 1.45	94.51 ± 0.35	88.11 ± 1.02	86.21 ± 0.84	94.86 ± 0.65
Libras	86.67 ± 0.79	89.33 ± 1.55	71.39 ± 1.37	71.78 ± 0.69	90.67 ± 1.24
ECG	83.80 ± 1.60	90.40 ± 1.11	80.50 ± 3.61	79.10 ± 4.13	91.40 ± 0.66
RobotFail	66.25 ± 1.88	61.56 ± 3.29	68.91 ± 1.77	58.28 ± 4.64	58.91 ± 2.10
CMU	100.0 ± 0.00	92.76 ± 3.60	98.97 ± 1.58	98.97 ± 1.58	97.59 ± 2.21
WALK	100.0 ± 0.00	100.0 ± 0.00	100.00 ± 0.00	100.0 ± 0.00	100.0 ± 0.00
KICK	50.23 ± 1.98	70.00 ± 0.00	67.00 ± 4.58	64.00 ± 11.14	81.00 ± 11.36
NetFlow	96.11 ± 0.66	93.01 ± 1.16	78.09 ± 0.00	91.74 ± 1.07	88.07 ± 1.06
Auslan	88.70 ± 1.38	97.76 ± 0.32	71.99 ± 1.54	77.14 ± 1.50	96.76 ± 0.70
Arab. Dig.	96.10 ± 0.35	99.13 ± 0.16	93.83 ± 0.45	91.99 ± 0.86	98.91 ± 0.19

Table 5.3 The accuracy rate values of the different models [29]/with permission of Elsevier.

Dataset	SMTS	TDVM	DD_{DWT}	CNN	LSTM-CNN	DESN
Jp. Vow.	96.5	96.6	98	97.66	—	98.13
Arab. Dig.	—	—	99.8	98.56	—	99.16
Ch.Traj.	—	—	99.1	98.25	100	98.11
ECG	86.6	—	85.5	86.04	86	86.04
Wafer	99	—	98.1	98.07	—	98.16
Auslan	97.3	—	81.5	72.83	97	78.48
CMU	99.3	—	96.3	93.81	100	95.00
Libras	—	—	95	93.61	97	94.72

Dataset	C_{ADE}	FESN	RC-RMS-SNP	RC-SNP	TSC-DR-NSNP	TSC-RMS-NSNP
Jp. Vow.	99.4	98.18	98.14	97.30	96.68	97.95
Arab. Dig.	99.2	99.23	99.13	93.83	91.99	98.91
Ch.Traj.	98	98.18	98.48	93.03	93.62	98.55
ECG	90	86.54	90.4	80.50	79.10	91.40
Wafer	98.3	98.07	99.2	98.46	98.50	99.27
Auslan	75.3	81.44	97.76	71.99	77.14	96.76
CMU	100	96.90	92.76	98.97	98.97	97.59
Libras	95.8	95.56	89.33	71.39	71.78	90.67

In the experiments, the discussed RC models are compared with eight baseline models. The evaluated approaches include two distance-based methods (DD_{DTW} [15] and TDVM [13], three feature extraction techniques (CNN [19], LSTM-FCN [17], and SMTS [16], along with three reservoir computing variants (FESN [21], C_{ADE} [20], and DESN). Table 5.3 displays accuracy comparisons among four reservoir computing variants and eight reference models tested on nine benchmark datasets. The baseline results and corresponding hyperparameter settings can be found in their original studies. Analysis indicates that: (i) RC-RMS-SNP outperforms all competitors on Auslan, Wafer, and ECG data; (ii) it achieves near-optimal accuracy for Jp. Vow, Arab. Dig., and Ch. Traj. data. Performance gaps remain, however, for uWave, Libras, CMU data where the model fails to match top accuracy levels.

References

1 M.M.A. Rahhal, Y. Bazi, H. AlHichri, N. Alajlan, F. Melgani, and R.R. Yager. Deep learning approach for active classification of electrocardiogram signals. *Information Sciences*, 345: 340–354, 2016.

2 A. Gharehbaghi and M. Lindén. A deep machine learning method for classifying cyclic time series of biological signals using time-growing neural network. *IEEE Transactions on Neural Networks and Learning Systems*, 29(9): 4102–4115, 2017.

3 J.A. Cortés-Ibáñez, S. González, J.J. Valle-Alonso, J. Luengo, S. García, and F. Herrera. Preprocessing methodology for time series: an industrial world application case study. *Information Sciences*, 514: 385–401, 2020.

4 A. Gupta, H.P. Gupta, B. Biswas, and T. Dutta. An early classification approach for multivariate time series of on-vehicle sensors in transportation. *IEEE Transactions on Intelligent Transportation Systems*, 21(12): 5316–5327, 2020.

5 S. Li, K. Li, and Y. Fu. Early recognition of 3D human actions. *ACM Transactions on Multimedia Computing, Communications, and Applications*, 14(1): 1–20, 2018.

6 A. Gupta, H.P. Gupta, B. Biswas, and T. Dutta. A fault-tolerant early classification approach for human activities using multivariate time series. *IEEE Transactions on Mobile Computing*, 20(5): 1747–1760, 2021.

7 Y. Wan and Y.-W. Si. A formal approach to chart patterns classification in financial time series. *Information Sciences*, 411: 151–175, 2017.

8 U. Mori, A. Mendiburu, I.M. Miranda, and J.A. Lozano. Early classification of time series using multi-objective optimization techniques. *Information Sciences*, 492: 204–218, 2019.

9 X. Xiao, R. Li, H.-T. Zheng, R. Ye, A. KumarSangaiah, and S. Xia. Novel dynamic multiple classification system for network traffic. *Information Sciences*, 479: 526–541, 2019.

10 W.A. Chaovalitwongse, R.S. Pottenger, S. Wang, Y.-J. Fan, and L.D. Iasemidis. Pattern-and network-based classification techniques for multichannel medical data signals to improve brain diagnosis. *IEEE Transactions on Systems Man and Cybernetics - Part A: Systems and Humans*, 41(5): 977–988, 2011.

11 E. Trentin, S. Scherer, and F. Schwenker. Emotion recognition from speech signals via a probabilistic echo-state network. *Pattern Recognition Letters*, 66: 4–12, 2015.

12 D. Hunt and D. Parry. Using echo state networks to classify unscripted, real-world punctual activity. *Engineering Applications of Neural Networks*. Cham, Switzerland: *Springer*, pages 369–378, 2015.

13 C. Orsenigo and C. Vercellis. Combining discrete SVM and fixed cardinality warping distances for multivariate time series classification. *Pattern Recognition*, 43(11): 3787–3794, 2010.

14 J. Mei, M. Liu, Y. Wang, and H. Gao. Learning a Mahalanobis distance based dynamic time warping measure for multivariate time series classification. *IEEE Transactions on Cybernetics*, 46(6): 1363–1374, 2016.

15 T. Górecki and M. Luczak. Multivariate time series classification with parametric derivative dynamic time warping. *Expert Systems with Applications*, 42(5): 2305–2312, 2015.

16 M.G. Baydogan and G. Runger. Learning a symbolic representation for multivariate time series classification. *Data Mining and Knowledge Discovery*, 29(2): 400–422, 2015.

17 F. Karim, S. Majumdar, H. Darabi, and S. Harford. Multivariate LSTM-FCNS for time series classification. *Neural Networks*, 116: 237–245, 2019.

18 A. Bagnall, J. Lines, J. Hills, and A. Bostrom. Time-series classification with COTE: the collective of transformation-based ensembles. *IEEE Transactions on Knowledge & Data Engineering*, 27(9): 2522–2535, 2015.

19 B. Zhao, H. Lu, S. Chen, J. Liu, and D. Wu. Convolutional neural networks for time series classification. *Journal of Systems Engineering and Electronics*, 28(1): 162–169, 2017.

20 L. Wang, Z. Wang, and S. Liu. An effective multivariate time series classification approach using echo state network and adaptive differential evolution algorithm. *Expert Systems with Applications*, 43: 237–249, 2016.

21 Q. Ma, L. Shen, W. Chen, J. Wang, J. Wei, and Z. Yu. Functional echo state network for time series classification. *Information Sciences*, 373: 1–20, 2016.

22 F.M. Bianchi, S. Scardapane, S. Løkse, and R. Jenssen. Reservoir computing approaches for representation and classification of multivariate time series. *IEEE Transactions on Neural Networks and Learning Systems*, 32(5): 2169–2179, 2021.

23 S. Hochreiter and J. Schmidhuber. Long short-term memory. *Neural Computation*, 9(8): 1735–1780, 1997.

24 K. Cho, B. Merrienboer, C. Gulcehre, D. Bahdanau, F. Bougares, H. Schwenk, and Y. Bengio. Learning phrase representations using RNN encoder-decoder for statistical machine translation. *arXiv:1406.1078*, 2014.

25 M. Lukovševičious and M. H. Jaeger. Reservoir computing approaches to recurrent neural network training. *Computer Science Review*, 3(3): 127–149, 2009.

26 H. Jaeger, M. Lukovševičious, D. Popovici, and U. Siewert. Optimization and application of echo state networks with leaky-integrator neurons. *Computer Science Review*, 20(3): 335–352, 2007.

27 H. Peng, Z. Lv, B. Li, X. Luo, J. Wang, X. Song, T. Wang, M.J. Pérez-Jiménez, and A. Riscos-Núñez. Nonlinear spiking neural P systems. *International Journal of Neural Systems*, 30(10): 2050008, 2020.

28 H. Peng, X. Xiong, M. Wu, J. Wang, Q. Yang, D. Orellana-Martín, and M.J. Pérez-Jiménez. Reservoir computing models based on spiking neural P systems for time series classification. *Neural Networks*, 169: 274–281, 2024.

29 X. Xiong, M. Wu, J. He, H. Peng, J. Wang, X. Long, and Q. Yang. Time series classification models based on nonlinear spiking neural P systems. *Engineering Applications of Artificial Intelligence*, 129: 107603, 2024.

6

Financial Time Series Analysis

6.1 Stock Market Index Prediction

6.1.1 Introduction

The inherent nonlinear complexity of financial systems makes stock market index forecasting a persistent research challenge that continues to engage scholars worldwide. In the early days of the study, researchers have proposed many classic prediction methods such as autoregressive, autoregressive and integrated moving average, and autoregressive and moving average. Traditional forecasting techniques often fail to capture the intricate patterns in financial time series data, as these datasets typically exhibit strong nonlinear characteristics and nonstationary behavior in real-world scenarios. In recent years, in the field of financial data prediction, machine learning models have become very popular, such as artificial neural networks (ANN) [1], support vector machines (SVM) [2], long short-term memory (LSTM) [3], gated recurrent unit (GRU) [4, 5], and recurrent neural network (RNN) [6].

This section focuses on how to apply NSNP systems to predict stock market index data. A stock market index forecasting method using NSNP systems will be discussed.

6.1.2 Models

6.1.2.1 K-order Nonlinear Spiking Neural P Systems

As a biologically-inspired computing paradigm, spiking neural P (SNP) systems emulate neuronal cell operations through membrane computing architectures. Their nonlinear extension, termed nonlinear spiking neural P (NSNP) systems, demonstrates enhanced computational capabilities. Comprehensive theoretical foundations for NSNP systems are established in Ref. [7]. For financial data processing, an enhanced version of NSNP systems with global weights is developed.

Spiking Neural P Systems for Time Series Analysis, First Edition. Jun Wang and Hong Peng.

Definition 6.1: A NSNP system with global weights is given by

$$\Pi = (O, \sigma_1, \sigma_2, \ldots, \sigma_n, syn, W)$$

wherein

1. $O = \{\alpha\}$ is an alphabet, where symbol α is called the spike;
2. $\sigma_1, \sigma_2, \ldots, \sigma_n$ are n neurons, with $\sigma_i = (v_i, r_i)$, $i = 1, 2, \ldots, n$, wherein
   a. $v_i \in \mathbb{R}$ is value of spikes in neuron σ_i;
   b. r_i is the spiking rule, of the form $\alpha^{g(v)} \rightarrow \alpha^{f(v)}$, where $g(\cdot)$ is called spike consumption function and $f(\cdot)$ is called spike generation function;
3. $syn \subseteq \{(i, j) | i, j = 1, 2, \ldots, n\}$; (synapse)
4. $W = (w_{ij})_{n \times n}$ is a synapse weight matrix, where w_{ij} is the weight of synapse from σ_j to σ_i.

The structure of a NSNP system is visually representable through a directed graph, where nodes stand for individual neurons and arrows depict synaptic pathways. The weight parameter w_{ij} assumes a nonzero value exclusively when a functional synapse exists from neuron σ_j to neuron σ_i; in all other cases, it defaults to null.

Let $v_i(\tau)$ denote the state of σ_i at timestep τ. Every neuron contains only one rule, $\alpha^{g(v)} \rightarrow \alpha^{f(v)}$. If neuron σ_i fires, then the spikes of $g(v_i(\tau))$ are removed, and the spikes of $f(v_i(\tau))$ are produced. Therefore, state of σ_i is updated by

$$v_i(\tau + 1) = v_i(\tau) - g(v_i(\tau)) + \sum_{j \neq i} w_{ij} f(v_j(\tau)) \tag{6.1}$$

wherein w_{ij} is the weight from σ_j to σ_i; $f(v_j(\tau))$ is value of spikes produced by σ_j at timestep τ. Let $g(v_i(\tau)) = v_i(\tau) - w_{ii} f(v_i(\tau))$. Then Eq. (6.1) is changed as

$$v_i(\tau + 1) = \sum_{j} w_{ij} f(v_j(\tau)) \tag{6.2}$$

or

$$v_i(\tau + 1) = w_{i1} f(v_1(\tau)) + \cdots + w_{ii} f(v_i(\tau)) + \cdots + w_{in} f(v_n(\tau)) \tag{6.3}$$

where w_{ij} is the weight parameter, $i, j = 1, 2, \ldots, n$.

In Eq. (6.3), the state of the system at time $\tau + 1$ depends only on the state at time τ. To depend on the states of the previous k times, Eq. (6.3) is given by

$$v_i(\tau + 1) = \sum_{j=1}^{n} [w_{ij}^1 f(v_j(\tau)) + w_{ij}^2 f(v_j(\tau - 1)) + \cdots + w_{ij}^k f(v_j(\tau - k)) + w_{i0}^0] \tag{6.4}$$

The mathematical expression (6.4) is designated as the state update equation for k-th order, while the resulting system is referred to as a k-order NSNP system.

By this convention, the system corresponding to Eq. (6.3) becomes a first-order NSNP system.

Functioning as a recurrent-type mechanism, the k-order NSNP system preserves information from k preceding states. Such systems demonstrate extensive dynamic variability. The current application focuses on employing this model to represent financial time series patterns.

When NSNP system is applied to model a dynamic system, it will encounter a difficulty: how to determine the appropriate number of neurons? The empirical mode decomposition will be adopted to solve this problem.

6.1.2.2 Empirical Mode Decomposition

As an adaptive signal processing technique introduced by Ref. [8], empirical mode decomposition (EMD) provides a unique approach for analyzing nonlinear and non-stationary signals. Unlike traditional decomposition methods including Fourier and wavelet transforms that rely on predefined basis functions, EMD adaptively separates signals based on their intrinsic temporal characteristics. It decomposes the series into a finite number of intrinsic mode functions (IMFs) and a residual. The decomposed intrinsic mode function components represent different characteristics of the original signal at different time scales.

EMD's theoretical framework requires: (i) presence of dual polarity peaks (maximum/minimum) in input signals; (ii) exclusive determination of signal behavior by inter-extrema durations; (iii) derivative-based extremum extraction when only inflection points exist, followed by integral reconstruction for complete decomposition.

An IMF must satisfy two key requirements: (i) across the complete data segment, the count of extremum points and zero-crossings should either match or differ by no more than one; (ii) the mean value between the upper envelope (defined by local maxima) and lower envelope (defined by local minima) must remain zero at all times, indicating local symmetry of both envelopes about the time axis.

Under these predefined assumptions and constraints, any given signal is separable into IMFs and a residual component. Theoretically, the EMD possesses universal applicability for time series analysis across all data types.

Following EMD decomposition, the initial signal $x(t)$ may be represented as

$$x(\tau) = \sum_{i=1}^{m} \mathrm{Imf}_i(\tau) + r_m(\tau) \tag{6.5}$$

wherein $\mathrm{Imf}_i(\cdot)$ denotes the ith IMF, and m is the number of total IMF; $r_m(\tau)$ is the final residual, which is either a constant or an average trend.

In this section, EMD is used as a signal processing tool. Its role is mainly reflected as follows: (i) it is used as a preprocessing and postprocessing tool; (ii) it can be used to determine the appropriate number of neurons for NSNP system.

6.1.3 Prediction Method

Let $x(1), x(2), \ldots, x(\tau)$ represent the historical data sequence. The objective involves employing NSNP systems to forecast the subsequent value $x(\tau + 1)$, which necessitates the design of an appropriate NSNP architecture. In the case of single-variable time series data, the one-dimensional numerical sequence cannot be immediately applied to form the multi-neuron architecture of an NSNP system. To address this, empirical mode decomposition (EMD) is employed to derive various temporal-scale characteristics from the initial series, producing either multiple subsequences or a multi-dimensional time series denoted as $V(1)$ through $V(\tau)$. These resulting multivariate sequences then serve as the foundation for constructing and training the NSNP network, with the neuron count corresponding precisely to the quantity of derived subsequences or decomposition levels. And then the well-trained NSNP system is used to predict the future values. Therefore, in addition to being a transform tool, another function of EMD is to determine the number of neurons in the automatically constructed NSNP system.

The constructed prediction model is called EMD-NSNP model. Figure 6.1 shows the overall structure of the EMD-NSNP model. The EMD-NSNP model is composed of two stages: (i) automatic construction and training (or learning) of NSNP system; (ii) generating the future value. In EMD-NSNP model, EMD is used as a preprocessing and postprocessing tool: (i) in the first stage, EMD is adopted to decompose the original time series into several time series in different resolution levels; (ii) in the second stage, inverse EMD is used to convert the values generated by the NSNP system as the future values of the original time series.

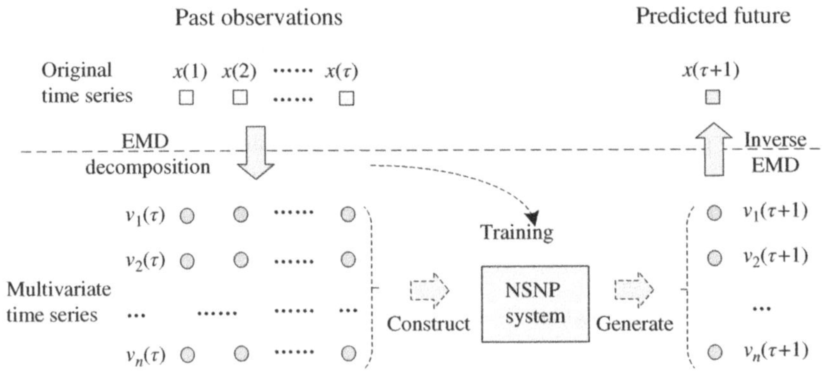

Figure 6.1 The structure of the EMD-NSNP model for stock index time series prediction (provided by authors).

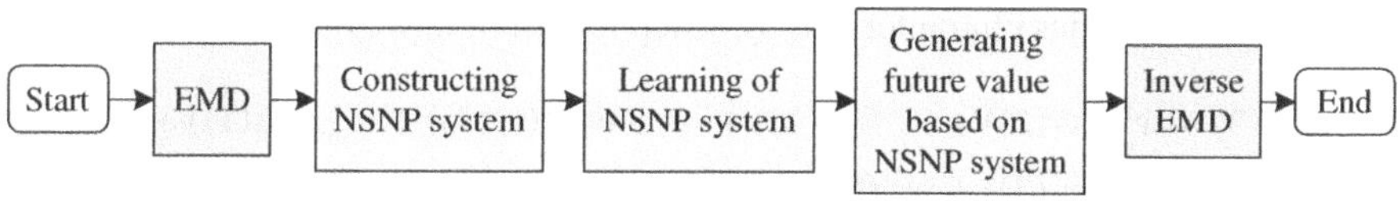

Figure 6.2 The block diagram of the EMD-NSNP model (provided by authors).

Figure 6.2 shows the block diagram of the EMD-NSNP model, where EMD is used as a preprocessing and post-processing tool. In the following, the three important components will be described in detail: constructing NSNP system, learning for NSNP system, and generating the future value.

6.1.3.1 Automatic Construction of NSNP System

First of all, original data, $x(1), x(2), \ldots, x(\tau)$, are usually nonlinear and nonstationary, therefore, the data must undergo min–max scaling to the range of $[-1, 1]$, as the EMD relies on zero-crossing points during its signal processing. Next, EMD is used to decompose the univariate time series into n multivariate time series at n resolution levels, resulting in a multivariate time series, $V(1), V(2), \ldots, V(\tau)$. The multivariate time series contains n sub-series: $v_1(1), v_1(2), \ldots, v_1(\tau); v_2(1), v_2(2), \ldots, v_2(\tau); \ldots; v_n(1), v_n(2), \ldots, v_n(\tau)$, where n is the number of decomposition series. That is, $V(\tau) = [(v_1(\tau), \ldots, v_n(\tau)]^T$.

Afterwards, a NSNP system consisting of n neurons is constructed automatically according to Definition 6.1. Particularly, if k is greater than 1, a k-order NSNP system can be constructed according to Eq. (6.4), that is, the constructed k-order NSNP system's state dynamics are governed by Eq. (6.4). Consequently, empirical mode decomposition serves to identify the optimal neuron count for this system.

6.1.3.2 Learning for NSNP System

The weight matrix of the constructed NSNP system can be trained according to the multivariate time series $V(1), V(2), \ldots, V(\tau)$, where $V(\tau) = [(v_1(\tau), \ldots, v_n(\tau)]^T$. A second-order NSNP framework serves as a case study to demonstrate the training methodology for implemented NSNP architectures.

The second-order NSNP system's state update formulation is derived from Eq. (6.4) as follows:

$$v_i(\tau + 1) = \sum_{j=1}^{n} [w_{ij}^1 f(v_j(\tau)) + w_{ij}^2 f(v_j(\tau - 1)) + w_{i0}^0] \tag{6.6}$$

wherein $i = 1, 2, \ldots, n$; the parameter n denotes both the neuron count in the NSNP system and the quantity of levels in empirical mode decomposition.

Given a time series characterized by a step length of L, we obtain

$$
\begin{aligned}
v_i(3) &= w_{ij}^1 f(v_1(2)) + w_{ij}^2 f(v_1(1)) + \cdots + w_{\text{in}}^1 f(v_n(2)) + w_{\text{in}}^2 f(v_n(1)) + w_{i0}^0 \\
v_i(4) &= w_{ij}^1 f(v_1(3)) + w_{ij}^2 f(v_1(2)) + \cdots + w_{\text{in}}^1 f(v_n(3)) + w_{\text{in}}^2 f(v_n(2)) + w_{i0}^0 \\
&\cdots\cdots \; \cdots\cdots \\
v_i(L) &= w_{ij}^1 f(v_1(L-1)) + w_{ij}^2 f(v_1(L-2)) + \cdots + w_{\text{in}}^1 f(v_n(L-1)) \\
&\quad + w_{\text{in}}^2 f(v_n(L-2)) + w_{i0}^0
\end{aligned}
\tag{6.7}
$$

Therefore, Eq. (6.7) can be written in the form of vector or matrix:

$$V_i = BW_i, \quad i = 1, 2, \ldots, n \tag{6.8}$$

where vectors V_i, W_i, and matrix B are, respectively, given by

$$V_i = [v_i(3), v_i(4), \ldots, v_i(L)]^T \tag{6.9}$$

$$W_i = [w_{i1}^1, w_{i1}^2, w_{i2}^1, w_{i2}^2, \ldots, w_{\text{in}}^1, w_{\text{in}}^2, w_{i0}^0]^T \tag{6.10}$$

$$
B = \begin{bmatrix}
f(v_1(2)) & f(v_1(1)) & \ldots & f(v_n(2)) & f(v_n(1)) & 1 \\
f(v_1(3)) & f(v_1(2)) & \ldots & f(v_n(3)) & f(v_n(2)) & 1 \\
\ldots & \ldots & \ldots & \ldots & \ldots & \ldots \\
f(v_1(L-1)) & f(v_1(L-2)) & \ldots & f(v_n(L-1)) & f(v_n(L-2)) & 1
\end{bmatrix}
\tag{6.11}
$$

For the ith neuron, its weight matrix is determined by solving the following problem:

$$\min_{w_i} ||V_i - BW_i||_2^2 + \sigma||W_i||_2 \tag{6.12}$$

where $||W_i||_2 = \left((w_{i1}^1)^2 + (w_{i1}^2)^2 + \cdots + (w_{\text{in}}^1)^2 + (w_{\text{in}}^2)^2 + (w_{i0})^2\right)^{1/2}$.

To solve Eq. (6.12), ridge regression method can be used, such as the Python library Scikit-learn.

6.1.3.3 Generating the Future Value

Following the learning phase, the optimized NSNP model produces predictive outputs, $V(\tau+1) = [(v_1(\tau+1), v_2(\tau+1), \ldots, v_n(\tau+1)]^T$, where $v_i(\tau+1)$ corresponds to the state of σ_i at timestep $\tau+1$.

Finally, the predicted value for original data at the next timestep, $x(\tau+1)$, can be obtained by inverse EMD transform according to $V(\tau+1)$.

6.1.4 Model Evaluation

In this section, experiments are done on seven major international stock market datasets to verify the forecasting performance of the EMD-NSNP method. Moreover, several baseline methods are used as the compared method to illustrate the superiority of the EMD-NSNP model.

6.1.5 Datasets

To evaluate the EMD-NSNP model, the daily closing prices of seven major international stock indices are collected, including S&P500 index (Standard & Poor's), NYSE Composite (NYSE), DJI (Dow Jones Industrial Average), NASDAQ Composite Index (NASDAQ Stock Exchange), FTSE100 Index (Financial Times Stock Exchange 100 Index), Nikkei 225 Index (Tokyo Stock Exchange) and SSE Index (Shanghai Stock Exchange). All prices are historical daily closing prices for a 9-year period from January 4, 2000 to December 31, 2018. These data comprising 15,723 financial records are collected from the open-access platforms of Yahoo Finance and Investing.org. Table 6.1 gives descriptive statistics for all indicators. These datasets can also be found in two open resources Investing.com[1] and Eastmoney.com[2].

6.1.6 Experimental Setup

First, before feeding the data into NSNP systems, it should be normalized to $[-1, 1]$, since EMD uses zero-crossing. As mentioned above, EMD is adaptive and

Table 6.1 Statistical information of seven datasets for stock market index (provided by authors).

Dataset	Study period	Size	Training	Testing
S&P500	01/04/2010–12/28/2018	2263	1810	453
NYSE	12/31/2009–12/28/2018	2264	1811	453
DJI	12/31/2009–12/28/2018	2264	1811	453
NASDAQ	12/31/2009–12/28/2018	2264	1811	453
FTSE	01/04/2010–12/31/2018	2272	1818	454
N225	01/04/2010–12/28/2018	2208	1766	442
SSE	01/04/2010–12/28/2018	2188	1750	438

1 https://cn.investing.com
2 https://choice.eastmoney.com

the number of nodes in NSNP systems does not need to be set manually. In automatic construction of NSNP systems, the optimal prior parameters are determined. To guarantee the objectivity of the results, the same division as the compared models are used. Each dataset is divided into two subsets: the first 80% are used as the training set, and the remaining 20% constitute the test set. Table 6.1 provides more details. To assess the performance of the EMD-NSNP model, four performance metrics are used in the experiments: mean squared error (MSE), mean absolute percentage error (MAPE), root mean squared error (RMSE), and mean absolute error (MAE). The metrics are defined by

$$\begin{aligned}
\text{MSE} &= \frac{1}{T}\sum_{\tau=1}^{T}(\hat{x}(\tau) - x(\tau))^2 \\
\text{RMSE} &= \sqrt{\frac{1}{T}\sum_{\tau=1}^{T}(\hat{x}(\tau) - x(\tau))^2} \\
\text{MAE} &= \frac{1}{T}\sum_{\tau=1}^{T}|\hat{x}(\tau) - x(\tau)| \\
\text{MAPE} &= \frac{1}{T}\sum_{\tau=1}^{T}\left|\frac{\hat{x}(\tau) - x(\tau)}{x(\tau)}\right| \times 100\%
\end{aligned} \tag{6.13}$$

wherein $x(\tau)$ and $\hat{x}(\tau)$ are the original and predicted values, and T is the number of samples in test set.

6.1.7 Experimental Results

6.1.7.1 Prediction Results for Seven Stock Market Indices

Table 6.2 provides the MSE, MAE, RMSE, and MAPE metric values of the EMD-NSNP method on seven stock market indices. Figure 6.3 shows the results of actual

Table 6.2 Four metric values of EMD-NSNP model (provided by authors).

Dataset	MSE	MAE	RMSE	MAPE
S&P500	326.85	14.65	18.07	0.56
NYSE	6110.09	56.94	78.16	0.46
DJI	93 127.5	256.67	305.16	1.09
NASDAQ	14 296.8	100.55	119.56	1.45
FTSE	1732.13	34.96	41.61	0.47
N225	20 756.8	114.55	144.07	0.53
SSE	823.36	23.74	28.69	0.79

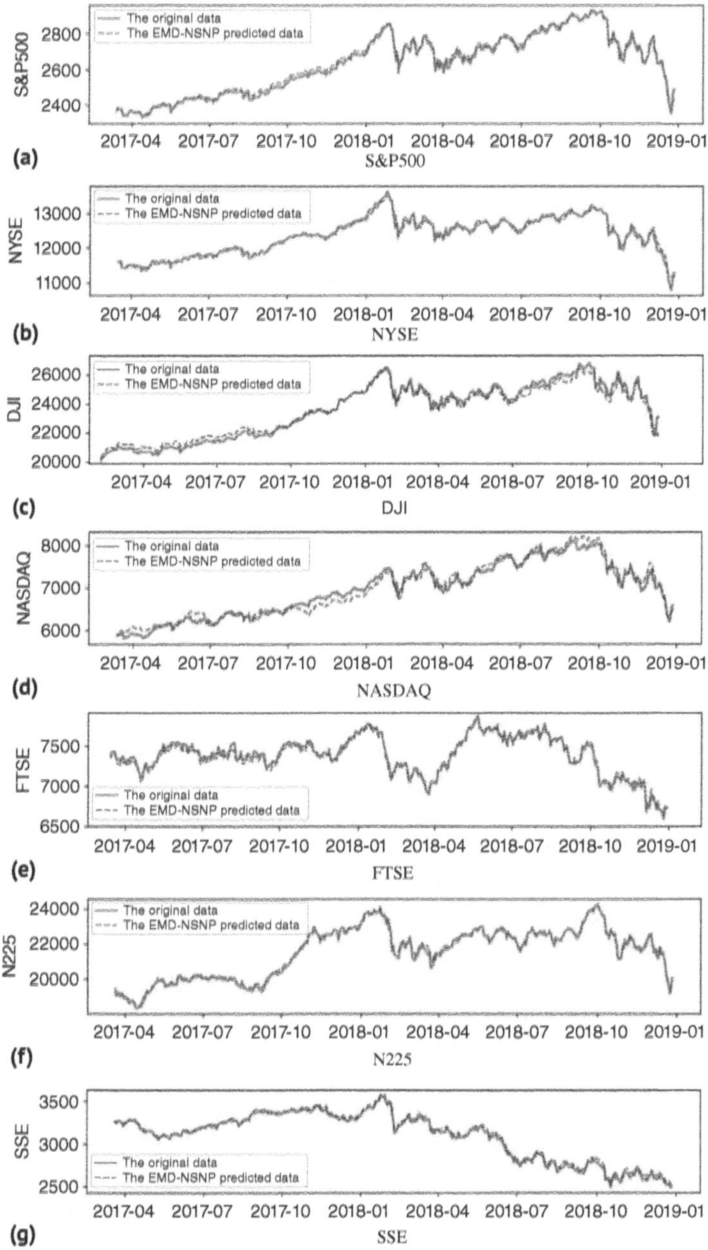

Figure 6.3 The prediction results of the EMD-NSNP model on seven stock index time series: (a) S&P500; (b) NYSE; (c) DJI; (d) NASDAQ; (e) FTSE; (f) N225; (g) SSE (provided by authors).

prices and the prices predicted by the EMD-NSNP model on seven stock index time series. Moreover, Figure 6.4 shows the absolute percentage errors of the EMD-NSNP model on seven stock index time series.

The EMD-NSNP model is compared with three baseline prediction models on seven stock index time series. The baseline models are RC, RNN, and LSTM. RC [9] is a reservoir computing model, while RNN and LSTM are two recurrent neural network models. The comparison results are provided in Table 6.3. It can be observed in Table 6.3 that, (i) for RMSE and MAE metrics, EMD-NSNP model outperforms the RC, RNN, and LSTM models on five datasets: S&P500, NYSE, FTSE, N225, and SSE; (ii) for RMSE and MAE metrics, RC achieves the best performance on NASDAQ dataset; (iii) for DJI dataset, RNN achieves the best prediction performance in terms of MAE, and RC achieves the best prediction performance in terms of RMSE. Evaluation outcomes reveal the EMD-NSNP framework's effectiveness for forecasting temporal sequences in stock market indices.

6.1.7.2 Comparative Study

To further compare the predictive ability of the EMD-NSNP model with several models reported in recent studies, two comparative experiments are conducted. To be able to directly compare, the same study period is used to train and test the EMD-NSNP model.

Case I: The EMD-NSNP model is compared with RC [9] and EMD2FNN [10] on S&P500, NASDAQ, and SSE datasets.

The research periods used in the datasets are: SSE (January 5, 2016 to December 30, 2016), NASDAQ (January 4, 2016 to December 30, 2016), and S&P500 (January 3, 2011 to December 30, 2011). Table 6.4 shows the comparisons of the EMD-NSNP with EMD2FNN and RC on S&P500, NAS-DAQ, and SSE. As shown in Table 6.4, the EMD-NSNP model outperforms EMD2FNN and RC models in predicting SP500, NASDAQ, and SSE indices, and the RMSE index drops by 21.97%, 10.89%, and 59.08%, respectively.

Case II: The EMD-NSNP model is compared with RC [9] models and ModAugNet [11] on S&P 500 and KOSPI200 data sets.

The trading period from January 2, 2008 to July 27, 2017 is used as the test set for S&P 500 and KOSPI200. Table 6.5 shows the comparison results of the EMD-NSNP model with ModAugNet and RC models on S&P 500 and KOSPI200 in terms of MAE, RMSE, and MAPE. As shown in Table 6.5, the EMD-NSNP model outperforms ModAugNet and RC models in predicting KOSPI200 index. For S&P 500 index, the EMD-NSNP is slightly worse than RC, and it is better than ModAugNet. The comparison results show that the EMD-NSNP is a promising stock market index forecasting model with competitive forecasting ability.

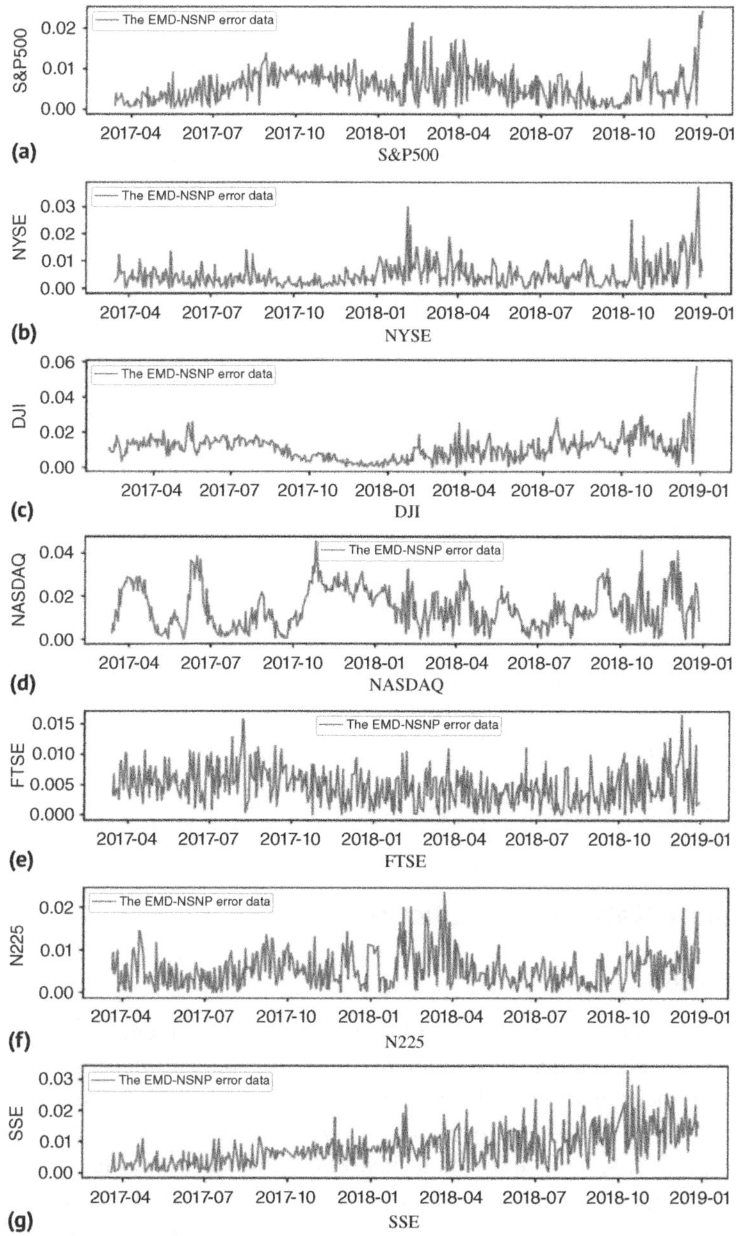

Figure 6.4 Absolute percentage errors (APE) of the EMD-NSNP model on seven stock index time series: (a) S&P500; (b) NYSE; (c) DJI; (d) NASDAQ; (e) FTSE; (f) N225; (g) SSE (provided by authors).

Table 6.3 Comparisons of the EMD-NSNP with RC, RNN, and LSTM on RMSE and MAE (provided by authors).

	RMSE			
Dataset	**EMD-NSNP**	**RC**	**RNN**	**LSTM**
SP500	8.07	23.25	23.44	23.89
NYSE	78.16	96.13	94.49	110.82
DJI	279.97	225.01	221.97	256.78
NASDAQ	119.56	78.09	78.80	81.41
FTSE	41.61	54.02	53.70	57.24
N225	144.07	229.35	229.92	267.24
SSE	28.69	30.76	30.89	33.08
	MAE			
Dataset	**EMD-NSNP**	**RC**	**RNN**	**LSTM**
SP500	14.65	15.80	15.37	15.66
NYSE	56.94	67.96	63.99	76.04
DJI	228.78	155.11	146.21	173.52
NASDAQ	100.55	53.24	53.80	56.11
FTSE	34.96	41.03	40.62	43.11
N225	114.55	160.77	163.33	190.58
SSE	21.74	22.07	22.08	23.99

6.2 Exchange Rate Price Prediction

6.2.1 Introduction

Financial markets exhibit significant volatility due to diverse influences, including economic cycles, geopolitical developments, and investor sentiment. Among the key areas of focus in financial research is time series analysis, particularly exchange rate prediction, which holds substantial importance for policymakers, corporations, and individual investors. Precise forecasts enable governments to design effective monetary policies, assist businesses in managing cross-border transaction risks, and guide private investors in optimizing portfolio returns. However, exchange rate dynamics form a complex, nonlinear system with

Table 6.4 Comparisons of the EMD-NSNP model with EMD2FNN and RC models on SP500, NASDAQ, and SSE (provided by authors).

Dataset	Model	MAE	RMSE	MAPE
SP500	EMD-NSNP	10.81	13.78	0.87
	EMD2FNN	13.04	17.66	1.05
	RC	13.59	18.00	1.10
NASDAQ	EMD-NSNP	34.00	43.87	0.66
	EMD2FNN	52.48	70.45	1.08
	RC	35.90	49.23	0.73
SSE	EMD-NSNP	13.92	17.48	0.43
	EMD2FNN	37.18	61.51	1.23
	RC	28.53	42.72	0.96

Table 6.5 Comparisons of the EMD-NSNP model with ModAugNet and RC models on SP500 and KOSPI200 (provided by authors).

Dataset	Model	MAE	RMSE	MAPE
SP500	EMD-NSNP	296.45	1.02	12.11
	ModAugNet	342.48	1.08	12.06
	RC	285.54	0.87	12.07
KOSPI200	EMD-NSNP	6.36	0.87	1.75
	ModAugNet	7.56	1.01	1.98
	RC	285.54	0.87	12.07

time-dependent variations, making accurate forecasting both a critical and demanding endeavor that continues to attract scholarly attention.

In this section, a recurrent-like model based on NSNP systems system is discussed for exchange rate forecasting task. First of all, a variant of NSNP systems is introduced, called gated nonlinear spiking neural P systems or GNSNP systems for short. Then, a unique GNSNP system is introduced, designed as a recurrent-like framework termed the GNSNP model. Building upon this structure, an enhanced recurrent model is developed specifically for exchange rate forecasting, named the ERF-GNSNP model [12]. To validate its performance, comprehensive experiments are carried out using real-world exchange rate datasets.

6.2.2 Model

6.2.2.1 Gated Nonlinear Spiking Neural P Systems

In this section, a variant of NSNP systems is introduced, gated nonlinear spiking neural P systems, termed as GNSNP systems. A key difference between GNSNP and NSNP systems lies in their incorporation of nonlinear gated structures. Additionally, GNSNP systems can be conveniently implemented as recurrent-like models.

Definition 6.2: A GNSNP system is a tuple

$$\Pi = (O, \sigma_1, \sigma_2, \ldots, \sigma_n, syn, in, out)$$

where

1. $O = \{\alpha\}$ is the singleton alphabet, in which α is known as the spike;
2. $\sigma_1, \sigma_2, \ldots, \sigma_n$ are n gated neurons, of the form $\sigma_i = (v_i, R_i, G_i)$, $i = 1, 2, \ldots, n$, where
   a. $v_i \in \mathbb{R}$ is the state of σ_i;
   b. R_i is a nonlinear spiking rule, with $\alpha^{g(v)} \to \alpha^{f(v)}$, in which $f(v)$ and $g(v)$ are linear function or nonlinear function;
   c. $G_i = \{r_i, c_i\}$, in which generation gate $c_i = \rho_2(\cdot)$ and reset gate $r_i = \rho_1(\cdot)$;
3. $syn \subseteq \{(i, j) | i, j = 1, 2, \ldots, n\}$; (synapse)
4. out and in, respectively, indicate output neuron and input neuron.

In GNSNP architectures, the framework is modeled using a directed graph structure, where gated neurons are interconnected via directed edges representing synaptic pathways. Every gated neuron has a data unit, which is used to express its internal state. Let $v_i(\tau)$ denote the state of gated neuron σ_i at time τ. Each gated neuron has the nonlinear spiking rule of the form $\alpha^{g(v)} \to \alpha^{f(v)}$, where $g(\cdot)$ is called spike consumption function and $f(\cdot)$ is called spike generation function.

When gated neuron σ_i fires at time τ, spiking rule $\alpha^{g(v)} \to \alpha^{f(v)}$ is applied. Consequently, $g(v)$ spikes are removed while $f(v)$ spikes are produced. As a result, the state update equation for the gated neuron σ_i can be formulated as:

$$v_i(\tau) = \rho_1(\tau) \cdot v_i(\tau - 1) - g(v) + m \tag{6.14}$$

where $v = x_i(\tau) + v_i(\tau - 1)$; $x_i(\tau)$ is external input at timestep τ, and $v_i(\tau - 1)$ and $v_i(\tau)$ denote the states of σ_i at timesteps $\tau - 1$ and τ; m is the spikes received from other neurons; $\rho_1(\tau)$ is the nonlinear function. Moreover, output equation is formulated as

$$h_i(\tau) = \rho_2(\tau) \cdot f(v) = \rho_2(\tau) \cdot f(x_i(\tau) + v_i(\tau)) \tag{6.15}$$

wherein the terms $v_i(\tau - 1)$ and $x_i(\tau)$ retain their previous definitions, $\rho_2(\tau)$ operates as a nonlinear gating mechanism, while $h_i(\tau)$ represents the output of the gated unit σ_i at timestep τ. The $\rho_1(\tau)$ and $\rho_2(\tau)$ are defined by

$$\rho_1(\tau) = \rho(W_r \cdot x(\tau) + V_r \cdot v(\tau - 1) + b_r) \tag{6.16}$$

$$\rho_2(\tau) = \rho(W_c \cdot x(\tau) + V_c \cdot v(\tau - 1) + b_c) \tag{6.17}$$

wherein W_c, V_c, W_r and V_r are matrices, and b_c and b_r are biases. In Eqs. (6.16) and (6.17), $\rho(\cdot)$ is an activation function, and often is set to be "hard_sigmoid". In Eq. (6.14), set $g(\cdot) \equiv 0$ and $m = 0$.

Figure 6.5 shows the function structure of GNSNP neuron, where $\mathbf{x}(\tau)$ is the input at τ; $\mathbf{v}(\tau - 1)$ and $\mathbf{v}(\tau)$ are the states for GNSNP neuron at $\tau - 1$ and τ; $\mathbf{h}(\tau)$ is its output at τ. The GNSNP neuron is realized as a recurrent-like model. Figure 6.6 shows an expansion of GNSNP neuron by time $\tau = 1, 2, \ldots, T$.

6.2.2.2 ERF-GNSNP Model

In this section, a recurrent-like model based on GNSNP neuron is designed for exchange rate forecasting tasks, termed as ERF-GNSNP model. In ERF-GNSNP model, GNSNP neuron is its key component. To realize the prediction of exchange

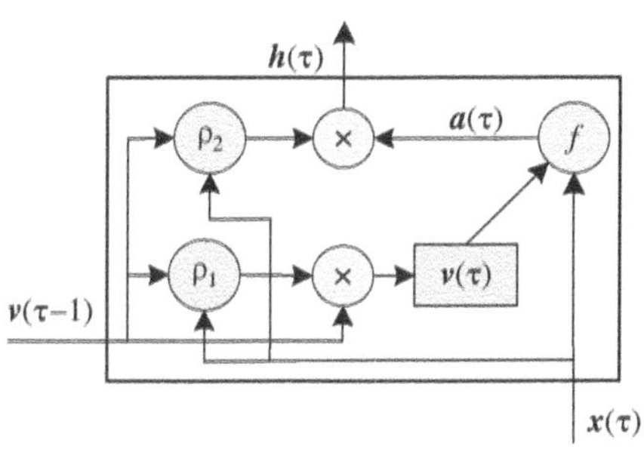

Figure 6.5 The neuron structure of GNSNP model [12]/World Scientific Publishing Co Pte Ltd.

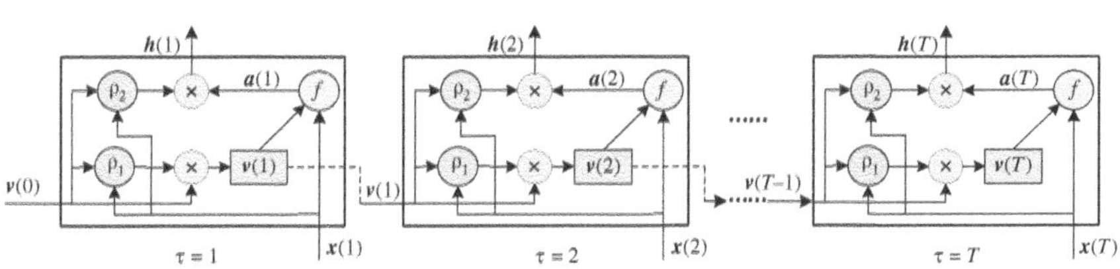

Figure 6.6 GNSNP model [12]/World Scientific Publishing Co Pte Ltd.

rate time series, GNSNP neuron is followed by a "dense" layer. For a multivariate time series, there is a certain correlation between its variables. The "dense" layer is used to capture the correlation, and it can be expressed by a linear model. Figure 6.7a shows the extended GNSNP neuron. Thus, the extended GNSNP neuron is formulated as

$$\begin{aligned}
\mathbf{v}(\tau) &= \rho_1(\tau) \odot \mathbf{v}(\tau - 1) \\
\mathbf{a}(\tau) &= f(W_a \cdot \mathbf{x}(\tau) + V_a \cdot \mathbf{v}(\tau) + b_a) \\
\mathbf{h}(\tau) &= \rho_2(\tau) \odot \mathbf{a}(\tau) \\
\mathbf{y}(\tau) &= W_y \cdot \mathbf{h}(\tau)
\end{aligned} \tag{6.18}$$

wherein $\rho_1(\tau), \rho_2(\tau)$ are given by Eqs. (6.16) and (6.17); $W_r, W_y, W_a, W_c, V_a, V_r, V_c, b_a, b_r$, and b_c are parameters to be learned.

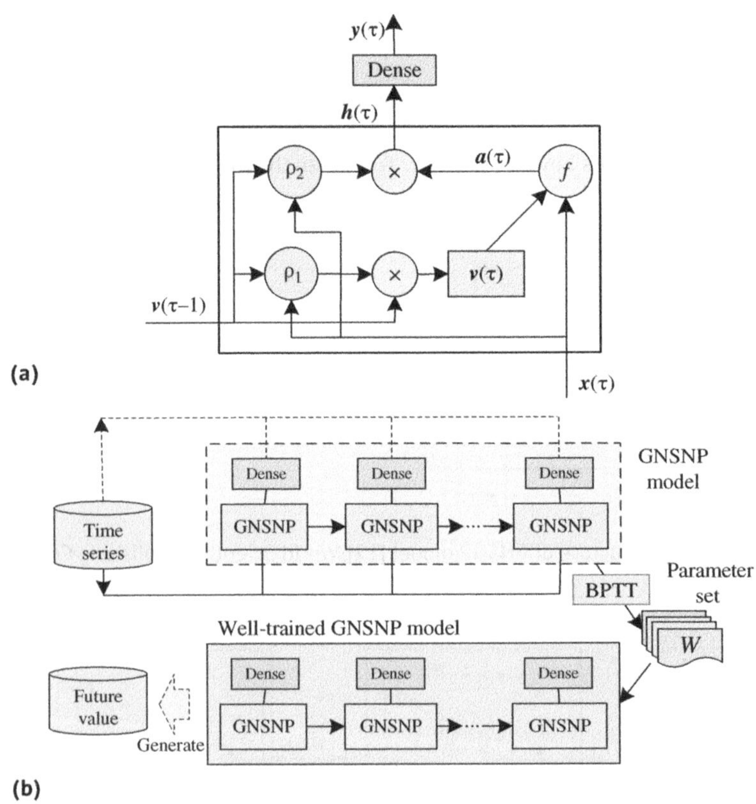

Figure 6.7 The ERF-GNSNP model: (a) an extended GNSNP neuron; (b) its two-stage implementation [12]/World Scientific Publishing Co Pte Ltd.

Developed within an RNN-based framework, the ERF-GNSNP model incorporates an enhanced GNSNP neuron. As depicted in Figure 6.7b, the method consists of two key steps. First, the BPTT technique fine-tunes the neuron and dense layer parameters. Following this, the refined GNSNP neuron is applied to predict upcoming values in exchange rate data.

6.2.3 Model Evaluation

In order to assess the ERF-GNSNP model's performance and applicability, ten currency pairs' historical price data are selected, with three experimental configurations implemented: two involving single-currency exchange rate timelines and one examining multi-currency correlated trends. In the experiments, four performance metrics are used: mean square error (MSE), root mean square error (RMSE), mean absolute error (MAE), and mean absolute percentage error (MAPE). These metrics are defined in Eq. (6.13).

6.2.3.1 Comparison on Univariate Exchange Rate Time Series

(A) Case Study I:

The case study selects four currency exchange rates against the RMB (CNY): GBP, JPY, EUR, and USD. The dataset spans from January 4, 2010, to February 28, 2018, as detailed in Table 6.6. For evaluation, the dataset is partitioned into training and test sets, with the last 396 samples reserved for testing and the remaining data used for training. The ERF-GNSNP model's performance on four univariate exchange rate time series is presented in Figure 6.8, showcasing both predicted values and associated errors. For every dataset, the left subplot contrasts raw data with model outputs, whereas the right subplot highlights prediction inaccuracies. It is evident that the model's estimates nearly overlap with the true values, demonstrating minimal discrepancies. This suggests that the ERF-GNSNP framework successfully learns the underlying trends in the exchange rate sequences.

To objectively evaluate the prediction performance of the ERF-GNSNP model, three baseline models (MLP, Elman, and SVR), two evolutionary SVR models

Table 6.6 Statistical information of exchange rate dataset in case study I [12]/World Scientific Publishing Co Pte Ltd.

Dataset	Study period	Size	Training set	Test set
GBPCNY	01/04/2010–02/28/2018	1981	1585	396
JPYCNY	01/04/2010–02/28/2018	1981	1585	396
EURCNY	01/04/2010–02/28/2018	1981	1585	396
USDCNY	01/04/2010–02/28/2018	1981	1585	396

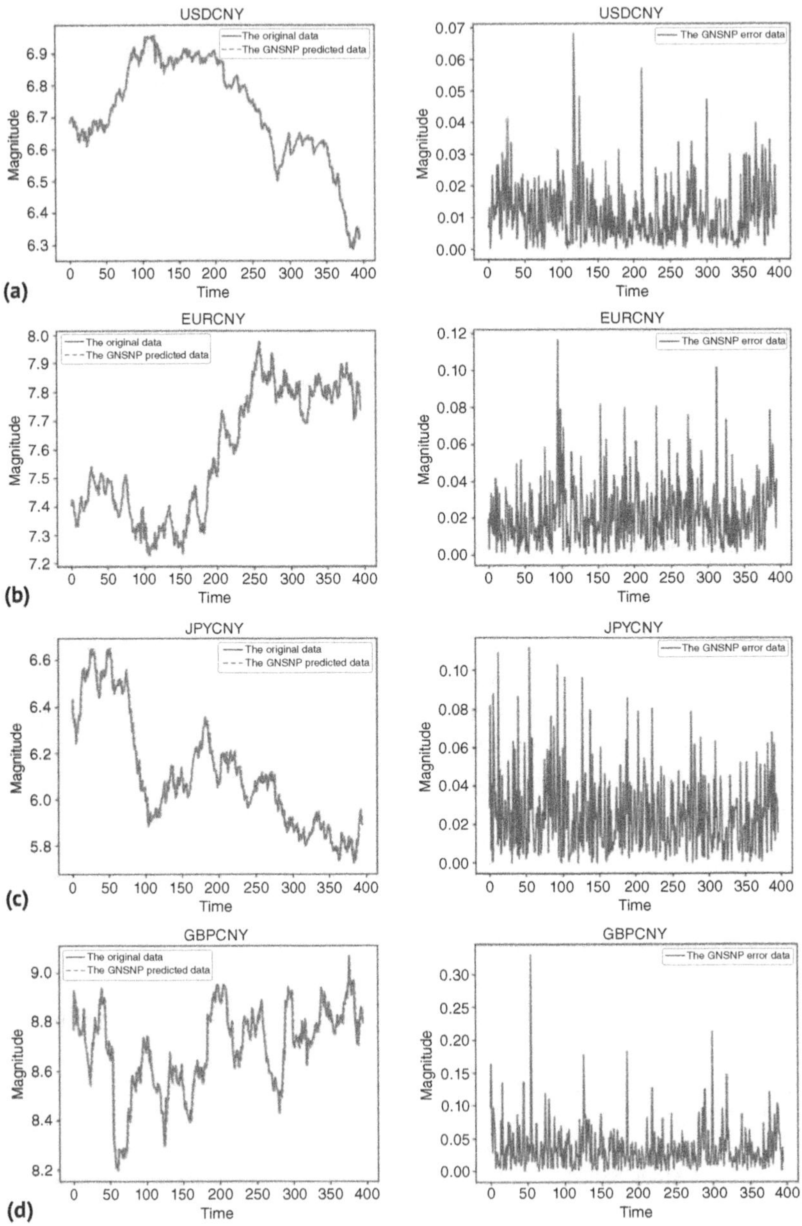

Figure 6.8 The predictive curves of the ERF-GNSNP model for USDCNY, EURCNY, JPYCNY, and GBPCNY data sets, where the left indicates original data curves and predicted data curves, and the right gives absolute prediction errors [12]/World Scientific Publishing Co Pte Ltd.

(GA-SVR and PSO-SVR) and GSNP are chosen to compare with it. Table 6.7 shows the comparison results, including RMSE, MAE and MAPE. The experimental results of MLP, Elman, SVR, GA-SVR, and PSO-SVR are retrieved from Ref. [13], and results of GSNP are reproduced according to its source code. Table 6.7 reveals the following findings: (i) The ERF-GNSNP model outperforms all benchmark models in RMSE, MAE, and MAPE for the USDCNY and GBPCNY datasets. (ii) Regarding the EURCNY dataset, ERF-GNSNP attains the lowest RMSE, while GSNP yields the best MAE and POS-SVR achieves the optimal MAPE. (iii) For JPYCNY, POS-SVR demonstrates superior MAPE performance, whereas GA-SVR excels in both RMSE and MAE.

Table 6.7 Comparisons of the ERF-GNSNP with other methods [12]/World Scientific Publishing Co Pte Ltd.

Model	Metric	USDCNY	EURCNY	JPYCNY	GBPCNY
	RMSE	0.0727	0.0418	0.0365	0.0902
MLPNN	MAE	0.0432	0.0226	0.0259	0.0685
	MAPE	0.6308	0.4348	0.424	0.7947
	RMSE	0.0197	0.0294	0.033	0.0709
Elman	MAE	0.0158	0.033	0.0283	0.0428
	MAPE	0.204	0.2973	0.4624	0.6491
	RMSE	0.0176	0.0288	0.0317	0.2845
CV-SVR	MAE	0.0137	0.0221	0.0271	0.2328
	MAPE	0.2346	0.2918	0.4033	2.7171
	RMSE	0.0155	0.0282	0.0313	0.0581
GA-SVR	MAE	0.0122	0.0219	0.0243	0.056
	MAPE	0.182	0.2885	0.4037	0.4967
	RMSE	0.0147	0.0281	0.0317	0.0504
POS-SVR	MAE	0.0114	0.0217	0.0247	0.0362
	MAPE	0.1707	0.2882	0.3972	0.4181
	RMSE	0.0148	0.0282	0.0325	0.0498
GSNP	MAE	0.0114	0.0215	0.0255	0.0365
	MAPE	0.1706	0.2885	0.4116	0.4022
	RMSE	0.0147	0.0280	0.0321	0.0489
ERF-GNSNP	MAE	0.0112	0.0219	0.0248	0.0348
	MAPE	0.1681	0.2898	0.4053	0.4014

(B) Case Study II:

The case study utilizes two currency exchange rate datasets: USD to Japanese Yen (USDJPY) and USD to RMB (USDCNY). The data spans from January 4, 2010, to December 29, 2017, with 87.5% allocated for training and the remaining 12.5% for testing. These datasets are sourced from Yahoo Finance and Investing.com. The descriptive statistics for the three datasets are provided in Table 6.8. Figure 6.9 displays the ERF-GNSNP model's forecasts and their corresponding errors across three exchange rate time series. For each series, the left subplot compares the observed and predicted values, whereas the right subplot visualizes the prediction errors. The results demonstrate that the model successfully follows the dynamic patterns of the exchange rates and achieves low prediction errors. This case study benchmarks the proposed GSNP against seven established models: ARMA, FFNN, two Lahmiri variants (EMD- and VMD-based RNNs), LSTM-U, GVL, and GVL-MB. Specifically, Lahmiri's methodologies merge mode decomposition with recurrent networks. LSTM-U operates as a univariate predictor, whereas GVL synergizes genetic optimization, VMD, and LSTM. The GVL-MB extension augments this architecture with BPNN components.

Table 6.9 provides the comparison results between the ERF-GNSNP model and nine baseline models (ARIMA model, SVR model, SVNN model, EMD-SVR model, EMD-SVNN model, VMD-SVR model, VMD-SVNN model, GVL-MB model, and GVL model) and GSNP on the exchange rate of US dollars to Japanese yen (US-DJPY) and the exchange rate of US dollars to RMB (USDCNY) datasets. Table 6.9 presents the MAPE scores for all evaluated models. The performance data for VMD-SVNN, VMD-SVR, EMD-SVNN, EMD-SVR, SVNN, SVR, and ARIMA are sourced from Ref. [14], while the results for GVL and GVL-MB originate from Ref. [15]. For GSNP, the outcomes are replicated based on its original implementation. As shown in the table, the ERF-GNSNP framework achieves significantly lower MAPE values compared to other models across both USDCNY and USDJPY datasets.

To conclude, both examples illustrate the strong predictive capability of the ERF-GNSNP framework for single-variable currency rate forecasting.

Table 6.8 Statistical information of exchange rate datasets in case study II [12]/World Scientific Publishing Co Pte Ltd.

Dataset	Study period	Size	Training set	Test set
USDCAD	01/02/2008–12/26/2013	1560	1248	312
USDCNY	01/04/2010–12/29/2017	2085	1824	261
USDJPY	01/04/2010–12/29/2017	2085	1824	261

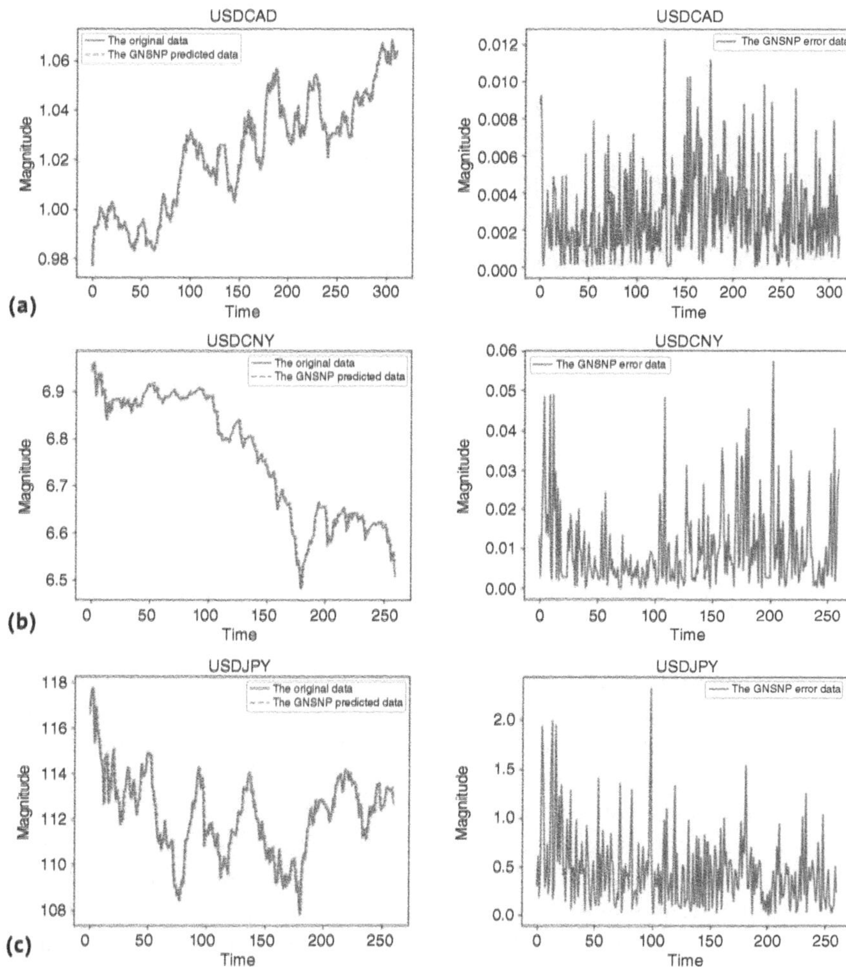

Figure 6.9 The predictive curves of the ERF-GNSNP model for USDCAD, USDCNY and USDJPY data sets, where the left indicates original data curves and predicted data curves, and the right gives absolute prediction errors [12]/World Scientific Publishing Co Pte Ltd.

6.2.3.2 Comparison on Multivariate Exchange Rate Time Series

The case study analyzes three currency pairs: EUR/USD, GBP/USD, and MXN/USD, using daily exchange rate data spanning from January 1, 2010, to December 31, 2019. Each dataset consists of 2,608 observations, recording opening, closing, high, and low prices. The data is split into training and testing subsets: the training set covers January 1, 2010, to December 29, 2017, while the test set includes data from January 1, 2018, to December 31, 2019.

Table 6.9 Comparisons of ERF-GN-SNP with other models for MAPE [12]/World Scientific Publishing Co Pte Ltd.

Model	USDCNY	USDJPY
GVL	0.169	0.406
GVL-MB	0.148	0.393
ARIMA	0.501	0.902
SVR	0.348	0.713
SVNN	0.354	0.645
EMD-SVR	0.293	0.523
EMD-SVNN	0.274	0.507
VMD-SVR	0.216	0.468
VMD-SVNN	0.197	0.406
GSNP	0.150	0.394
ERF-GNSNP	0.147	0.391

The ERF-GNSNP model's performance on three multivariate exchange rate test sets is visualized in Figure 6.10. Each pair of subplots contrasts the original data with the model's predictions (left) and their respective errors (right). The near-perfect overlap between predicted and actual curves, along with negligible error magnitudes, confirms the model's exceptional ability to replicate the underlying patterns in the exchange rate time series.

To assess the performance of the ERF-GNSNP model, it is evaluated against 11 benchmark methods using three multivariate time series datasets. The competing models consist of GSNP, ALS-RS, LSTM-RS, RS, ANN-GARCH, RF, Adaboost, Bagging, SVR, DT, and ARIMA. Among these, ALS-RS employs a novel ensemble learning strategy that leverages the synergy between shallow and deep features through an adaptive linear sparse random subspace framework. Further descriptions of these models are available in prior research (see Ref. [16]. The comparative analysis, presented in Table 6.10, includes MAPE and MSE metrics. While results for ALS-RS, LSTM-RS, RS, ANN-GARCH, RF, Adaboost, Bagging, SVR, DT, and ARIMA are sourced from Ref. [16], the GSNP outcomes are replicated based on its original implementation. The results in Table 6.10 indicate that: (i) For MXNUSD, ERF-GNSNP provides the best accuracy in terms of MAPE and MSE; (ii) With EURUSD and GBPUSD, ERF-GNSNP dominates in MSE, while ALS-RS performs better in MAPE. Such outcomes highlight the robustness of the ERF-GNSNP method for multivariate exchange rate prediction.

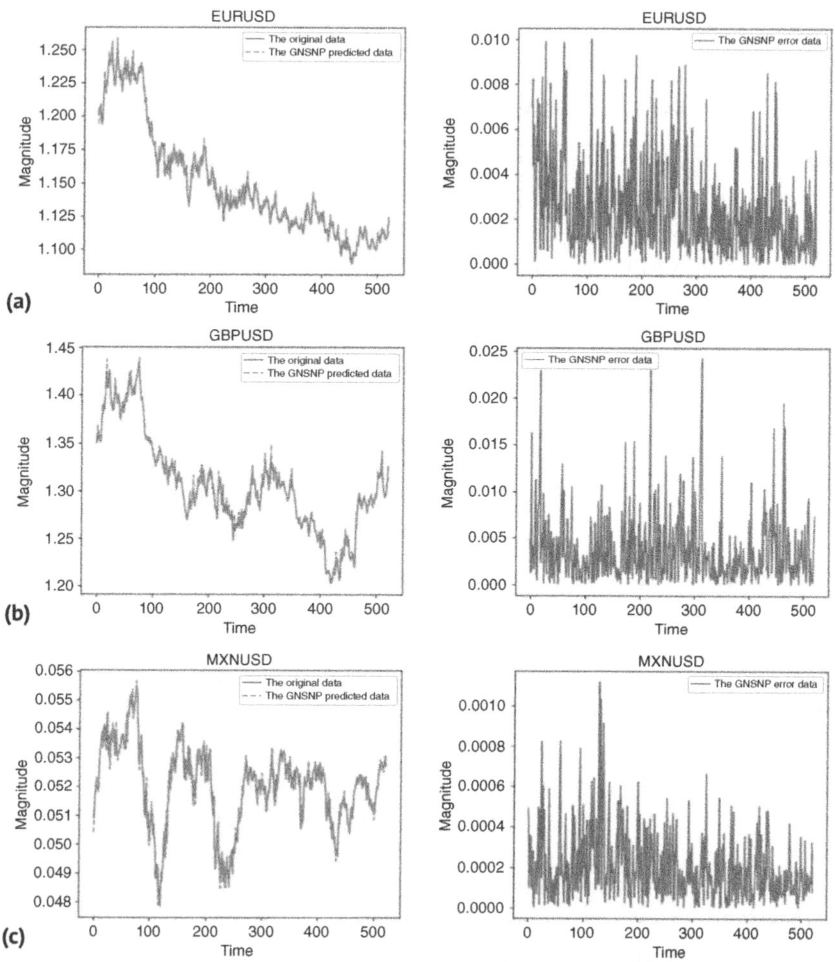

Figure 6.10 The predictive curves of the ERF-GNSNP model for EURUSD, GBPUSD and MXNUSD data sets, where the left indicates original data curves and predicted data curves, and the right gives absolute prediction errors [12]/World Scientific Publishing Co Pte Ltd.

6.3 Crude Oil Price Prediction

6.3.1 Introduction

The global energy sector heavily relies on crude oil, whose market fluctuations directly influence international economic stability. Dramatic price swings may impair economic growth trajectories and create security challenges for nations

Table 6.10 Comparisons of ERF-GNSNP with other models [12]/World Scientific Publishing Co Pte Ltd.

Model	Metric	EURUSD	GBPUSD	MXNUSD
ARIMA	MSE	0.05672	0.07941	0.04504
	MAPE	0.468	0.7365	0.7459
DT	MSE	0.02067	0.01273	0.03590
	MAPE	0.4643	0.6231	0.7341
Adaboost	MSE	0.01624	0.01217	0.007
	MAPE	0.2718	0.2434	0.0547
Bagging	MSE	0.01622	0.0121	0.00771
	MAPE	0.2681	0.243	0.0612
SVR	MSE	0.01734	0.01231	0.01175
	MAPE	0.3284	0.5267	0.0728
RF	MSE	0.00231	0.00851	0.02242
	MAPE	0.2635	0.3452	0.5156
ANN-GARCH	MSE	0.00162	0.01242	0.00844
	MAPE	0.2657	0.5306	0.0653
RS	MSE	0.00043	0.00659	0.00686
	MAPE	0.1273	0.1913	0.0541
LSTM-RS	MSE	0.00027	0.00593	0.00584
	MAPE	0.1015	0.1349	0.0535
ALS-RS	MSE	0.0002	0.0059	0.00552
	MAPE	0.0936	0.1138	0.0524
GSNP	MSE	0.0015	0.0039	0.00521
	MAPE	0.1907	0.2554	0.0368
ERF-GNSNP	MSE	0.0001	0.0020	0.0007
	MAPE	0.1860	0.2551	0.0332

dependent on stable energy supplies. However, crude oil price fluctuations are primarily governed by supply–demand dynamics in conjunction with movements in related commodity markets [17], constituting the core pricing mechanism. In addition, crude oil price is still affected by many uncertain factors, for example, extreme events [18], political stability [19], economic growth [20], etc. The crude oil prices present the complex features of non-stationary and nonlinear, thus, it is difficult to predict crude oil prices. Therefore, forecasting crude oil futures prices with precision and efficiency presents a significant challenge.

In recent years, to deal with the limitations of traditional prediction models, many deep learning and artificial intelligence approaches have been used to crude oil price prediction, for example SVM [21], ANN [22], etc. These methods were applied to solve the nonlinear problem of crude oil prices. Although various machine learning models were used to predict crude oil price and some of them have shown good performance, crude oil price time series prediction is still a challenging task because of the complex nonlinear dynamic characteristics of the corresponding systems.

In this section, a variant of echo state network (ESN) is introduced, where the reservoir is an NSNP-TO system, called echo-like NSNP system or ENSNP system. Based on the ENSNP system, a new prediction model is discussed for international crude oil price forecasting task, called AENSNP model.

6.3.2 Model

In subsection 4.4.1, NSNP-TO system is discussed as a variant of NSNP system. In this section, the NSNP-TO system is used to construct a variant of echo state network (ESN), called echo-type NSNP system or termed as ENSNP system. The ENSNP system has an input layer, a NSNP-TO system and an output layer, that is, a NSNP-TO system, is equipped with an output layer and an input layer. Figure 6.11a shows the ENSNP system.

Structurally, an ESNP system functions as an ESN adaptation, employing an NSNP-TO system for reservoir computation. Thus, the ENSNP systems can be realized in the framework of ESN.

Thc purpose of designing ENSNP systems is to construct a forecasting method for international crude oil prices. In many sequence-based tasks, the attention mechanism can catch the most relevant features in the input and make more accurate predictions. Thus, a trainable soft attention mechanism is introduced before the ENSNP system processes sequence input. Therefore, the predictive model is called AENSNP model. Figure 6.11b shows the AENSNP model. The AENSNP model contains two components: (i) Soft attention component: original input $\mathbf{x}(\tau) = (x_1(\tau), \dots, x_m(\tau))^T$ is processed by soft attention mechanism to obtain s input features $\mathbf{z}(\tau) = (z_1(\tau), \dots, z_s(\tau))^T$; (ii) An ENSNP system, which is composed of a feature input layer, an NSNP-TO system, and an output layer. The feature input layer has s input units. The NSNP-TO system provides nonlinear dynamics for the ENSNP system. The output layer has k units, and the output at τ is expressed by $\mathbf{y}(\tau) = (y_1(\tau), \dots, y_k(\tau))^T$. In addition, regularization constraint is used to enhance the generalization capability of the model. The AENSNP model is implemented in the framework of ESN. Thus, as usual in ESN, the trainable parameters of soft attention component and NSNP-TO systems are initialized randomly, and other trainable parameters are trained by ridge regression algorithm.

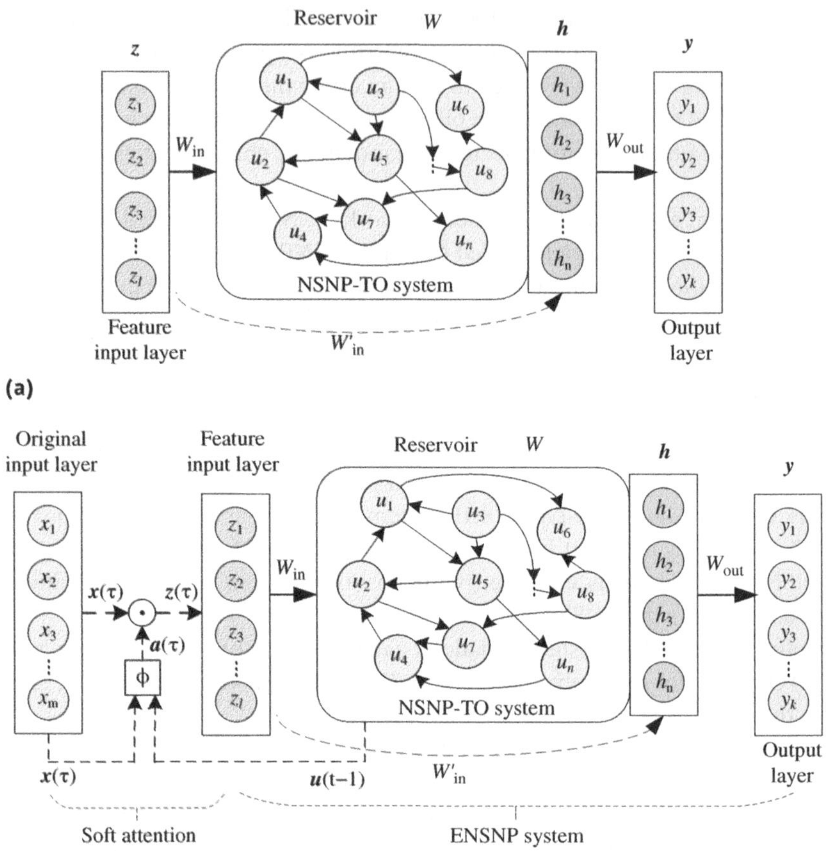

Figure 6.11 (a) An ENSNP system; (b) AENSNP model (provided by authors).

In the process of model training, the soft attention mechanism responds through $\mathbf{a}(\tau)$, and $\mathbf{a}(\tau)$ is a vector. The $\mathbf{a}(\tau)$ is computed by

$$\mathbf{a}(\tau) = \phi(\beta_1 \mathbf{v}_i(\tau - 1) + W_a f(W_{\text{ina}} \mathbf{x}(\tau) + \beta_2 \mathbf{v}(\tau - 1))) \tag{6.19}$$

where β_1 and β_2 are two scalar parameters; W_{a} and W_{ina} are trainable parameters in soft attention mechanism; $\phi(\cdot)$ is an activation function, and it is set to be softmax; $\mathbf{v}(\tau - 1)$ is the state of the reservoir (or NSNP-TO system) at $\tau - 1$, and $\mathbf{x}(\tau)$ is the input at τ. The reservoir state $\mathbf{v}(\tau - 1)$ is used to determine the important data points in the input $\mathbf{x}(\tau)$. The soft attention mechanism focuses on the input characteristics of saliency through element-level multiplication. Thus, the soft attention mechanism is denoted by

$$\mathbf{z}(\tau) = \mathbf{x}(\tau) \odot \mathbf{a}(\tau) \tag{6.20}$$

where symbol $\odot$ denotes element-level multiplication. After the soft attention, the computation of ENSNP system depends on the feature input $\mathbf{z}(\tau)$ rather than the original input $\mathbf{x}(\tau)$.

Based on Eq. (6.20), state and output equations for AENSNP model are defined by

$$\begin{aligned} \mathbf{a}(\tau) &= \phi(\beta_1 \mathbf{v}_i(\tau - 1) + W_a f(W_{\text{ina}}\mathbf{x}(\tau) + \beta_2 \mathbf{v}(\tau - 1))) \\ \mathbf{z}(\tau) &= \mathbf{x}(\tau) \odot \mathbf{a}(\tau) \\ \mathbf{v}(\tau) &= \beta_1 \mathbf{v}(\tau - 1) + Wf(W_{\text{in}}\mathbf{z}(\tau) + \beta_2 \mathbf{v}(\tau - 1)) \\ \mathbf{h}(\tau) &= f_o(\gamma \mathbf{v}_i(W'_{\text{in}}\mathbf{z}(\tau) + \tau - 1)) \\ \mathbf{y}(\tau) &= W_{\text{out}}\mathbf{h}(\tau) \end{aligned} \tag{6.21}$$

wherein $\mathbf{y}(\tau) = (y_1(\tau), \ldots, y_k(\tau))^T$ denotes the output at τ; $\beta_1, \beta_2, \gamma \in [0, 1]$ are three scalar parameters, and $W, W_a \in \mathbb{R}^{n\times n}$, $W_{\text{in}}, W_{\text{ina}} \in \mathbb{R}^{m\times n}$, $W'_{\text{in}} \in \mathbb{R}^{m\times n}$ and $W_{\text{out}} \in \mathbb{R}^{n\times k}$ are six parameter matrices; f is sigmoid function; f_o is linear function or nonlinear function, typically, it is an identity function.

Since AENSNP model (or ENSNP system) uses a structure similar to ESN, the same supervised learning method can be adopted for training it. The training of AENSNP includes two stages: initialization and training periods. In initialization period, five matrices W, W_a, W_{in}, W_{ina}, and W'_{in} can be initialized randomly. In training period, weigh matrix W_{out} is determined by ridge regression algorithm.

Suppose that $\mathbf{x}(\tau) = (x_1(\tau), \ldots, x_m(\tau))^T$ is the original input at timestep τ, and the output of NSNP-TO system is $\mathbf{h}(\tau) = (h_1(\tau), \ldots, h_n(\tau))^T$. Thus, the output of AENSNP model, $\mathbf{y}(\tau) = (y_1(\tau), \ldots, y_k(\tau))^T$, can be calculated by

$$\mathbf{y}(\tau) = W_{\text{out}}\mathbf{h}(\tau) \tag{6.22}$$

Suppose that we have L original input data, $\mathbf{x}(1), \ldots, \mathbf{x}(L)$, and the output of NSNP-TO system is $\mathbf{h}(1), \ldots, \mathbf{h}(L)$ and the output of AENSNP model is $\hat{\mathbf{y}}(1), \ldots, \hat{\mathbf{y}}(L)$. Let $H = [\mathbf{h}(1), \ldots, \mathbf{h}(L)]$ and $Y = [\hat{\mathbf{y}}(1), \ldots, \hat{\mathbf{y}}(L)]$. Then, we can compute

$$Y = W_{\text{out}}H \tag{6.23}$$

Therefore, W_{out} can be computed by

$$W_{\text{out}} = (H^T H)^{-1} HY \tag{6.24}$$

where $(H^T H)^{-1}$ denotes the inverse of $H^T H$, However, to avert the ill-posed problems caused by the matrix when its dimension becomes larger, W_{out} is solved by

$$W_{\text{out}} = (H^T H + \lambda I)^{-1} HY \tag{6.25}$$

where $I \in \mathbb{R}^{n \times n}$ denotes unit matrix, and λ is a regularization factor. In implementing the AENSNP model, output matrix W_{out} can be obtained by ridge regression algorithm to solve the following problem:

$$W_{out}^* = \underset{W_{out}}{\text{argmin}} J(W_{out}) = \frac{1}{2}||Y - W_{out}H||^2 + \frac{1}{2}\lambda||W_{out}||^2 \tag{6.26}$$

6.3.3 Model Evaluation

To validate the availability and advantages of the AENSNP method in crude oil price forecasting, two crude oil data from major crude oil futures market are chosen as the benchmark data sets: London International Petroleum Exchange and New York Mercantile Exchange's West Texas Intermediate crude oil (WTI crude oil) and North Sea Brent crude oil (Brent crude oil). Three groups of experiments are designed: (i) daily, weekly, and monthly Brent and WTI crude oil time series; (ii) Brent and WTI crude oil data in three different time periods; (iii) Daily Brent and WTI crude oil price. That is, empirical analysis on 14 crude oil price data sets is completed. In the experiments, before sending the sequence data to the method, each data is normalized to [−1, 1].

Seven performance metrics are used: root mean square error (RMSE), mean absolute error (MAE), mean square prediction error ratio (MSPE), mean absolute scale error (MASE), mean square error (MSE), standard deviation of error (SDE), and mean absolute percentage error (MAPE). The MAPE, MSE, MAE, and RMSE metrics are defined in Eq. (6.13). The SDE, MASE, and MSPE metrics are given by:

$$\begin{aligned}
\text{SDE} &= \sqrt{\frac{1}{T}\sum_{\tau=1}^{T}\left(\hat{x}(\tau) - x(\tau) - \frac{1}{N}\sum_{\tau=1}^{T}(x(\tau) - \hat{x}(\tau))\right)^2} \\
\text{MASE} &= \frac{1}{T}\frac{\sum_{\tau=2}^{T}|x(\tau) - \hat{x}(\tau)|}{\frac{1}{T-1}\sum_{\tau=1}^{T}|x(\tau) - \hat{x}(\tau)|} \\
\text{MSPE} &= \frac{1}{T}\sqrt{\sum_{\tau=1}^{T}\left(\frac{x(\tau) - \hat{x}(\tau)}{x(\tau)}\right)^2}
\end{aligned} \tag{6.27}$$

where T denotes the number of testing samples, and $\hat{x}(\tau)$ and $x(\tau)$ are the predicted value and the original value.

6.3.3.1 Monthly, Weekly, and Daily Crude Oil Price Forecasting

The first group of experiments uses monthly, weekly, and daily Brent and WTI crude oil spot price datasets. The six datasets are retrieved from the official website of the United States Energy Information Administration (https://www.eia.gov/). These data vary from January 1, 2010 to December 31, 2019. In order to study the AENSNP model, each dataset is partitioned as two parts, including training and test sets. In this group of experiments, the first 80% of samples are used as the training set (January 1, 2010 to December 21, 2017), and the last 20% are used as the test set (January 31, 2018 to December 30, 2019). Table 6.11 shows the description of these data sets.

By comparing with nine different baseline prediction methods, the prediction performance of the AENSNP model is studied. The comparison methods include five benchmark models (Naive model, ARMA model, BPNN model, ENN model, and SVM model), two composite methods (VMD+SVM and CEEMDAN+SVM), and two multi-scale methods (VMD+SVM+ARMA and CEEMDAN+SVM+AMMA). More details of the above comprised models can be found in Ref. [23]. Figures 6.12 and 6.13 show the prediction values and prediction errors of the AENSNP model for six time series of Brent and WTI (daily, weekly and monthly). For each data set, the left figure gives the prediction curve and original curve, while the right figure gives their absolute error curves. It can be seen that the AENSNP model can well catch the change trends of WTI and Brent in four time series of daily and weekly, and the prediction error is small. However, the fitting effect of monthly WTI and Brent data is slightly worse than that of daily and weekly forecasts.

Tables 6.12 and 6.13 show the comparisons of the AENSNP model with nine prediction methods. The experimental results of the comparison models are sourced from Ref. [23]. It can be seen from Tables 6.12 and 6.13 that in the above six data

Table 6.11 Statistical description of crude oil spot price data in experiment 1 (provided by authors).

Datasets		Size	Training set	Test set
WTI	WTI_monthly	120	96	24
	WTI_weekly	522	418	104
	WTI_daily	2513	2014	499
Brent	Brent_monthly	120	96	24
	Brent_weekly	522	418	104
	Brent_daily	2530	2021	509

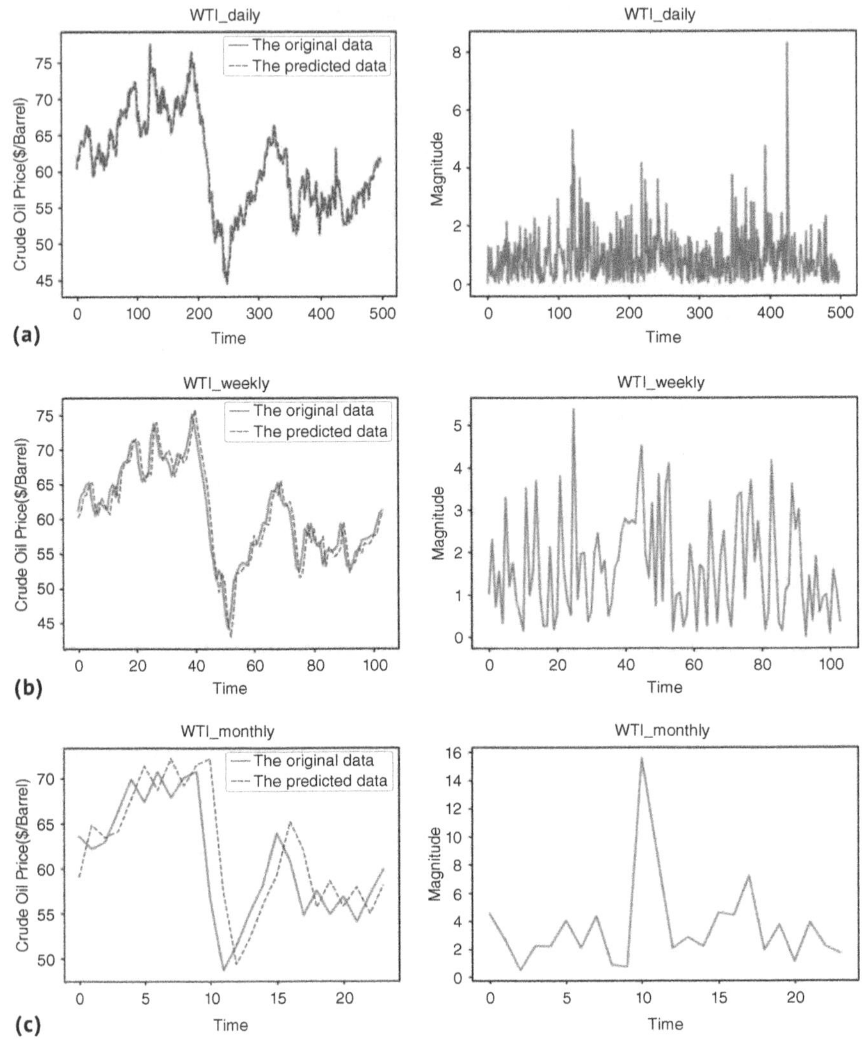

Figure 6.12 The daily, weekly, and monthly crude oil price prediction results of the AENSNP model for the WTI in Experiment 1 (provided by authors).

sets, the MSE and SDE index values of the AEN-SNP method are superior to other nine methods, and have significant lower values, where for WTI weekly's, its MSE and SDE are 0.033 and 0.043, respectively, which are 91.20% and 92.89% lower than VMD+SVM+ARMA model. However, for the MASE index, it is also noticed that the AENSNP model performs worse in predicting WTI and Brent crude oil prices, which is lower than VMD+SVM and VMD+SVM+ARMA methods.

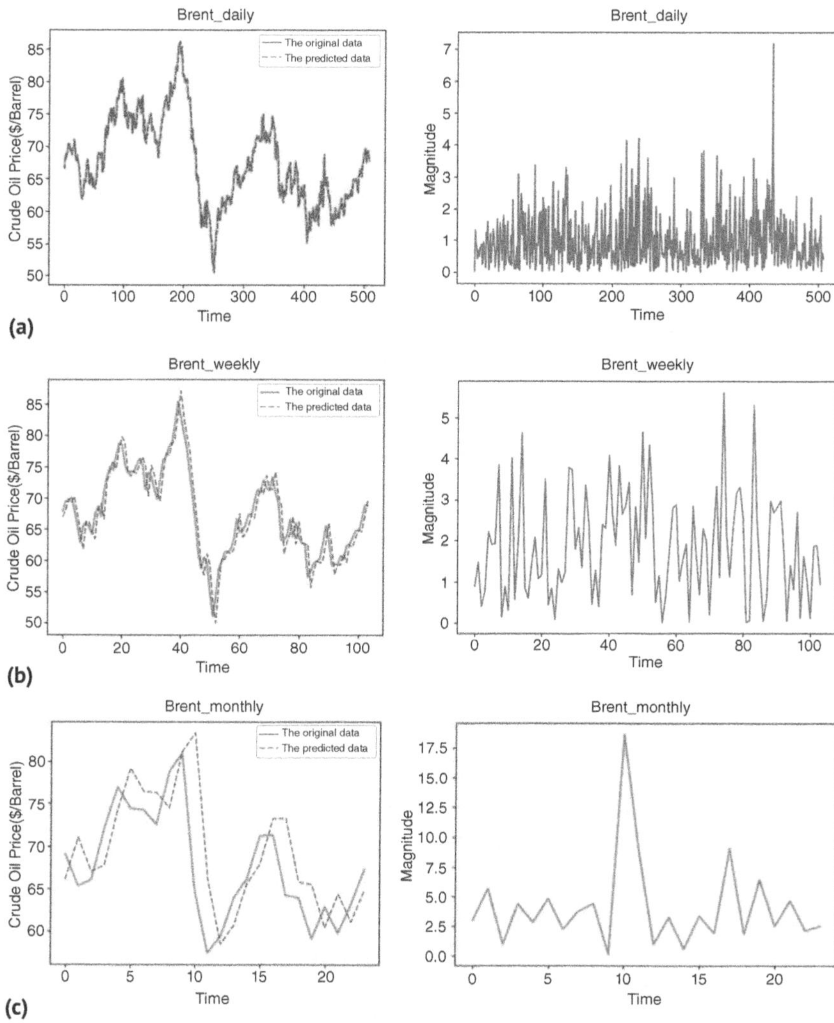

Figure 6.13 The daily, weekly, and monthly crude oil price prediction results of the AENSNP model for the Brent in Experiment 1 (provided by authors).

6.3.3.2 Different Time Periods Daily Crude Oil Price Forecasting

In the second experiment, six data sets of the daily spot prices of Brent and WTI crude oil in three different time periods are selected. These data sets are from the official website of the United States Energy Information Administration (https://www.eia.gov/). In each sequence data, the first 70% is used as training samples, while the remaining 30% constitutes test samples. The statistical description of these six data sets is shown in Table 6.14.

Table 6.12 Comparison results of different prediction methods on WTI crude oil price (provided by authors).

Model	Metric	WTI_daily	WTI_weekly	WTI_monthly
Naive	MSE	10.964	21.785	345.307
	SDE	3.287	4.667	16.823
	MASE	2.067	2.418	3.012
ARMA	MSE	1.564	4.077	20.771
	SDE	1.249	2.004	4.401
	MASE	1.031	0.778	0.184
BPNN	MSE	1.600	9.778	75.706
	SDE	1.257	2.128	6.889
	MASE	1.060	1.168	0.928
ENN	MSE	3.203	5.110	32.246
	SDE	1.753	2.260	5.679
	MASE	1.164	0.932	0.851
SVM	MSE	1.562	4.178	25.083
	SDE	1.249	2.042	4.904
	MASE	1.054	0.805	1.219
VMD+SVM	MSE	1.253	2.130	18.659
	SDE	1.118	1.458	4.316
	MASE	0.982	0.649	0.891
CEEMDAN+SVM	MSE	1.755	2.680	8.011
	SDE	1.317	1.637	2.827
	MASE	1.319	0.692	0.116
VMD+SVM+ARMA	MSE	0.188	0.375	2.802
	SDE	0.428	0.605	1.673
	MASE	0.463	0.321	0.370
CEEMDAN+SVM+ARMA	MSE	3.474	3.868	11.465
	SDE	1.225	1.595	2.173
	MASE	1.437	0.933	0.141
AENSNP	MSE	0.024	0.033	0.200
	SDE	0.028	0.043	0.131
	MASE	1.004	1.014	1.099

Table 6.13 Comparison results of different prediction methods on Brent crude oil price (provided by authors).

Model	Metric	WTI_daily	WTI_weekly	WTI_monthly
Naive	MSE	9.106	357.074	247.438
	SDE	2.978	17.509	14.994
	MASE	1.458	0.877	0.291
ARMA	MSE	1.712	5.017	27.379
	SDE	1.307	2.221	5.085
	MASE	0.990	0.279	0.134
BPNN	MSE	1.772	5.396	81.856
	SDE	1.329	2.309	8.751
	MASE	0.966	0.282	0.196
ENN	MSE	3.003	6.263	40.821
	SDE	1.731	2.482	6.319
	MASE	1.243	0.325	0.161
SVM	MSE	1.858	4.885	55.627
	SDE	1.362	2.203	6.716
	MASE	0.939	0.786	1.441
VMD+SVM	MSE	1.221	2.601	35.924
	SDE	1.105	1.613	5.845
	MASE	0.922	0.228	0.163
CEEMDAN+SVM	MSE	1.475	2.819	16.707
	SDE	1.214	1.670	4.056
	MASE	1.057	0.240	0.124
VMD+SVM+ARMA	MSE	0.200	0.485	6.334
	SDE	0.438	0.686	2.304
	MASE	0.431	0.113	0.083
CEEMDAN+SVM+ARMA	MSE	3.476	13.646	9.174
	SDE	1.383	2.209	2.563
	MASE	1.112	0.414	0.065
AENSNP	MSE	0.066	0.0581	0.408
	SDE	0.025	0.045	0.112
	MASE	1.010	1.007	1.022

Table 6.14 Statistical description of crude oil spot price data in experiment 2 (provided by authors).

Oil	Dataset ID	Training set	Test set	Sample capacity
WTI	1	1998.1–2004.12	2005.1–2008.1	2503
	2	2002.1–2009.12	2010.1–2013.1	2762
	3	2009.1–2016.12	2017.1–2020.1	2765
Brent	1	1998.1–2004.12	2005.1–2008.1	2548
	2	2002.1–2009.12	2010.1–2013.1	2790
	3	2009.1–2016.12	2017.1–2020.1	2782

ARIMA, AR, and TV-TD models are adopted to compare the presented models. ARIMA is classical forecasting model, while TV-TD is a time-varying trend decomposition model. The details of the above models can be found in Ref. [24]. Figures 6.14 and 6.15 show the predictive results of the AENSNP model for WTI and Brent crude oil in three different time periods. For every data set, the left figure gives the predictive curve and original curve, while the right figure gives their absolute error curves. It can be seen that the AENSNP model can well catch the change trends of each time series, and the predictive errors are small.

Table 6.15 provides the results of the AENSNP method and five methods on WTI and Brent data sets in three different time periods. The experimental results of the comparison models are sourced from Ref. [24]. It can be seen from Table 6.15 that in the three datasets, the AENSNP achieves very good prediction results. For MSPE value, in data sets 1, 2, and 3, the predicted value of AENSNP model is significantly lower than that of other methods. The minimum MSPE index is the Brent crude oil data set (datasets2-Brent) in data set 2, with a value of 0.0331, which is 95.91% lower than the second best method. For the MAPE index, datasets2-Brent has the best prediction effect in the three data sets, with a value of 0.2293, which is 74.68% lower than the second best method.

6.3.3.3 Daily Crude Oil Price Prediction

The third group of experiments use the original prices of Brent and WTI crude oil futures, whose data is from investing website (https://cn.investing.com). The daily closing price of Brent and WTI crude oil futures varies from 2010-01-04 to 2020-07-31, with data length of 2,746 and 2,732, respectively. Each sequence is partitioned into two parts, with the first 70% of the sequence as training set and the last 30% as testing set. In order to evaluate the prediction performance, 12 different methods are selected to compare the AENSNP method. These comparison methods include seven benchmark models (CNN, LSTM, GRU, BP, SVR, BLAC, and BiLSTM models) and five combined methods (BiLSTM-CNN, CNN-BiLSTM,

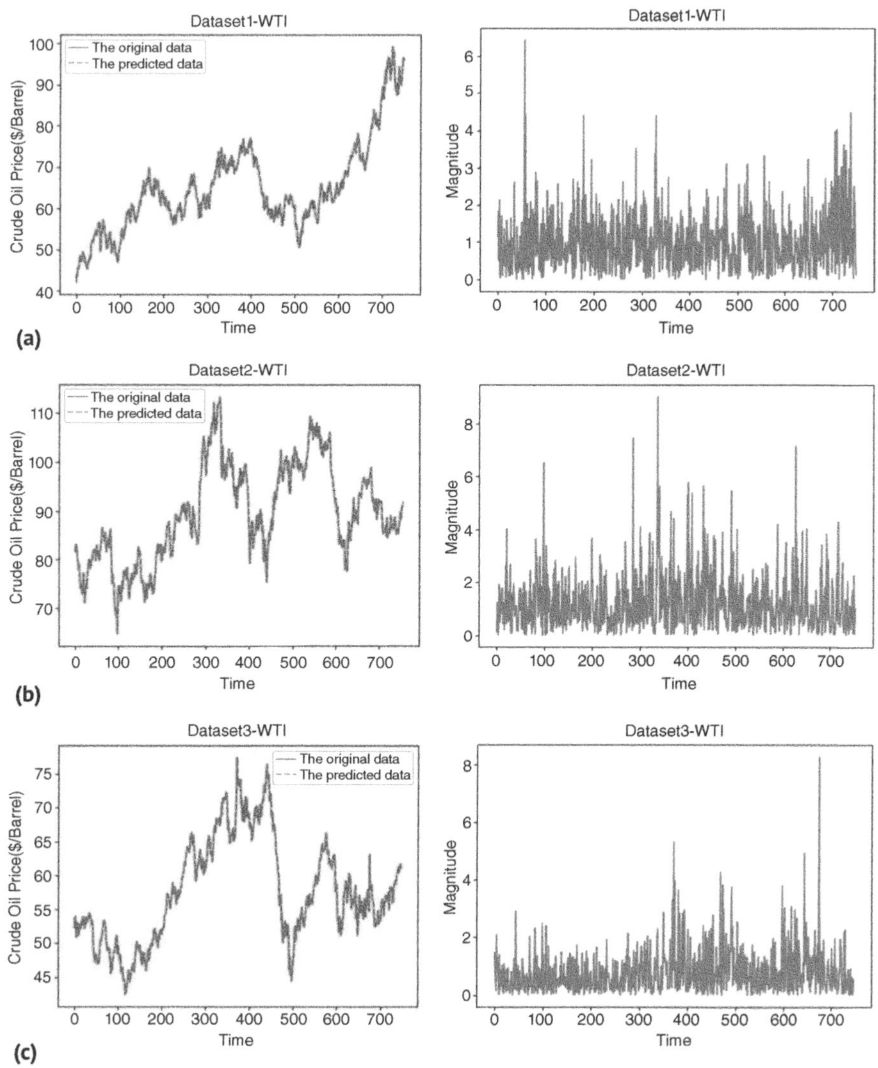

Figure 6.14 The prediction results of the AENSNP model for WTI crude oil prices in three different periods in Experiment 2 (provided by authors).

LSTM-CNN, CNN-LSTM, and WT-BLAC models). More details of these comparison methods can be found in Ref. [25].

Figure 6.16 shows the prediction results of the AENSNP method in Brent and WTI crude oil data sets. The left figure gives the original curve and prediction curve, while the right figure gives their absolute error curves. It can be seen that

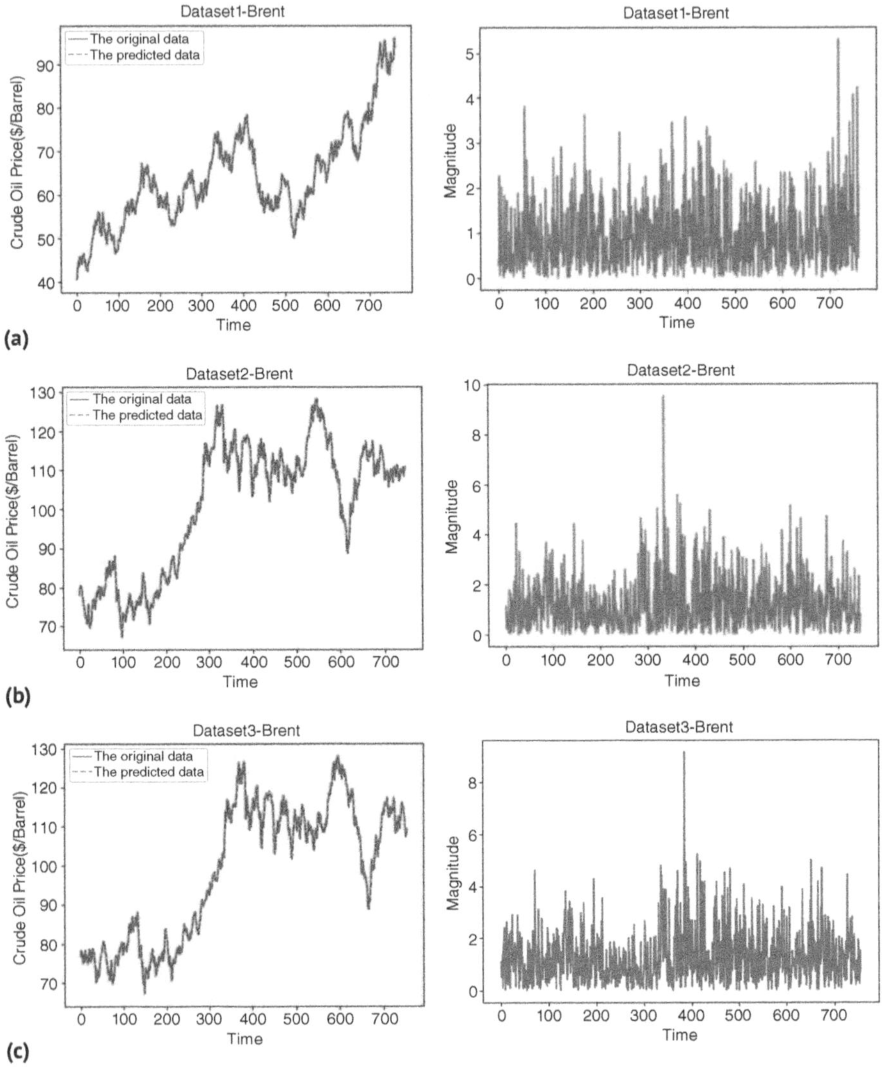

Figure 6.15 The prediction results of the AENSNP model for Brent crude oil prices in three different periods in Experiment 2 (provided by authors).

the AENSNP model can well catch the change trends of each time series, and the predictive errors are small.

Table 6.16 provides the results of the AENSNP model and 12 baseline models on WTI and Brent data sets. The experimental results of the comparison models are retrieved from Ref. [25]. We can clearly see from Table 6.16 that the AENSNP

Table 6.15 Comparison results of different prediction methods on WTI and Brent crude oil prices (provided by authors).

Datasets		Metrics	TV-TD	ARIMA (1,1,0)	ARIMA (1,2,1)	ARIMA (2,1,0)	AENSNP
Dataset 1	WTI	MAPE	1.0615	0.9394	0.9386	0.9406	0.2735
		MSPE	1.1190	0.8179	0.8157	0.8176	0.0347
	Brent	MAPE	0.9768	0.9991	1.1511	0.9993	0.2737
		MSPE	1.1548	0.9990	1.3255	0.9999	0.0435
Dataset 2	WTI	MAPE	0.9541	1.0010	1.0021	1.0011	0.6406
		MSPE	0.8886	1.0016	1.0033	1.0024	0.1477
	Brent	MAPE	0.9057	0.9992	1.0004	0.9990	0.2293
		MSPE	0.8099	0.9998	1.0037	1.0006	0.0331
Dataset 3	WTI	MAPE	0.8900	0.9997	1.0010	1.0009	0.2883
		MSPE	0.8154	0.9971	1.0003	0.9986	0.0470
	Brent	MAPE	0.9992	1.0026	1.0050	1.0023	0.8022
		MSPE	0.9427	1.0040	1.0077	1.0039	0.5531

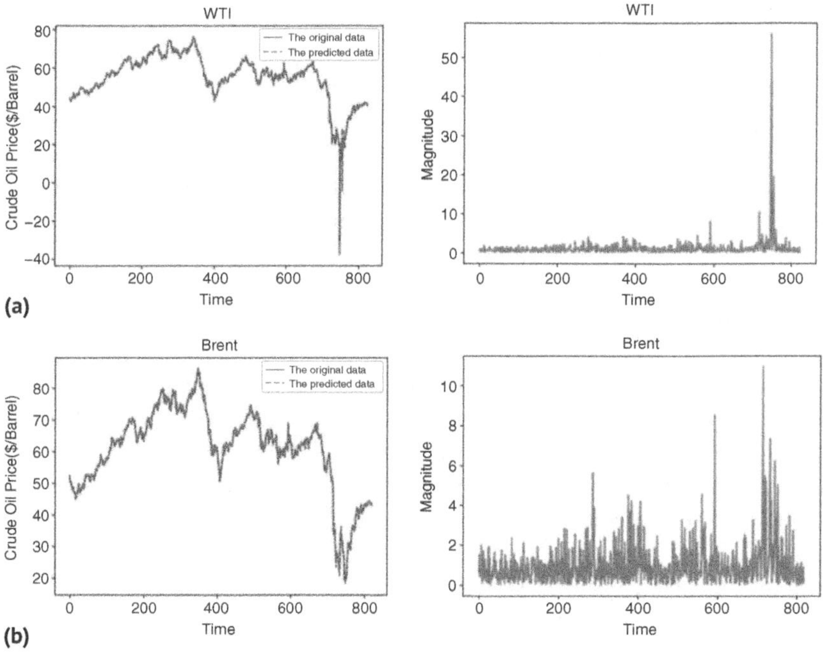

Figure 6.16 The prediction results of AENSNP for the Brent and WTI crude oil prices in Experiment 3 (provided by authors).

Table 6.16 Comparison results of different prediction methods on Brent crude oil price (provided by authors).

	WTI			Brent		
Model	**RMSE**	**MAE**	**MAPE**	**RMSE**	**MAE**	**MAPE**
BiLSTM	2.6928	1.2849	2.9770	1.7186	1.3041	2.3963
BLAC	2.6181	1.1656	2.7294	1.6336	1.1777	2.2150
CNN-LSTM	4.1226	2.2733	6.1430	3.7575	2.1827	5.2557
CNN	5.3265	4.1991	9.1029	4.0692	2.7261	6.2662
CNN-BiLSTM	4.8437	2.4861	7.3427	3.5752	2.2134	5.1565
GRU	2.8336	1.4153	3.2947	1.7657	1.2726	2.4771
LSTM	2.9619	1.5572	3.6216	2.1963	1.6489	3.2725
BiLSTM-CNN	3.4235	2.4637	4.8848	1.9098	1.2681	2.6778
LSTM-CNN	3.6368	2.4733	5.4053	2.1606	1.5404	3.0925
BP	2.9653	1.4142	3.5639	2.1809	1.5196	3.1735
SVR	5.6715	3.3630	9.3963	5.6269	4.8044	9.7157
WT-BLAC	2.2518	1.1780	2.6261	1.2936	0.8894	1.8027
AENSNP	0.0351	0.0159	0.1590	0.0722	0.1198	0.8834

model has the highest fit for the forecast of Brent and WTI crude oil future prices, and its MAPE, MAE, and RMSE are the smallest of all methods. Among them, the MAPE, MAE, and RMSE of AENSNP in WTI are 0.0351, 0.0159, and 0.1590, respectively, which are 98.44%, 98.65%, and 93.94% lower than WT-BLAC model. For this experiment, we also see that the AENSNP model is obviously better than these recurrent neural networks, such as GRU, LSTM, and BiLSTM. These results show that the AENSNP model is very advantageous for predicting oil prices.

References

1 M. Paluch and L. Jackowska-Strumillo. Hybrid models combining technical and fractal analysis with ANN for short-term prediction of close values on the Warsaw stock exchange. *Applied Sciences-Basel*, 8(12): 2473, 2018.

2 M.-C. Lee. Using support vector machine with a hybrid feature selection method to the stock trend prediction. *Expert Systems with Applications*, 36: 10896–10904, 2009.

3 S. Borovkova and I. Tsiamas. An ensemble of LSTM neural networks for high frequency stock market classification. *Journal of Forecasting*, 38: 600–619, 2019.

4 Q. Liu, L. Long, H. Peng, J. Wang, Q. Yang, X. Song, A. Riscos-Núñez, and M.J. Pérenz-Jiménez. Gated spiking neural P systems for time series forecasting. *IEEE Transactions on Neural Networks and Learning Systems*, 34(9): 6227–6236, 2023.

5 Q. Liu, H. Peng, L. Long, J. Wang, Q. Yang, M.J. Pérez-Jiménez, and D. Orellana-Martín. Nonlinear spiking neural systems with autapses for predicting chaotic time series. *IEEE Transactions on Cybernetics*, 54(3): 1841–1853, 2024.

6 T. Gao and Y. Chai. Improving stock closing price prediction using recurrent neural network and technical indicators. *Neural Computing*, 30: 2833–2854, 2018.

7 H. Peng, Z. Lv, B. Li, X. Luo, J. Wang, X. Song, T. Wang, M.J. Pérez-Jiménez, and A. Riscos-Núñez. Nonlinear spiking neural P systems. *International Journal of Neural Systems*, 30(10): 2050008, 2020.

8 N. Huang, Z. Shen, S. Long, M. Wu, H. Shih, Q. Zheng, N. Yen, C. Tung, and H. Liu. The empirical mode decomposition and the Hilbert spectrum for nonlinear and non-stationary time series analysis. *Proceedings of the Royal Society A, Mathematical Physical and Engineering Sciences*, pages 903–995, 1998.

9 W. Wang, Y. Tang, J. Xiong, and Y. Zhang. Stock market index prediction based on reservoir computing models. *Expert Systems with Applications*, 178: 115022, 2021.

10 F. Zhou, M. Zhou, Z. Yang, and L. Yang. EMD2FNN: a strategy combining empirical mode decomposition and factorization machine based neural network for stock market trend prediction. *Expert Systems with Applications*, 115: 136–151, 2019.

11 Y. Baek and H.Y. Kim. Modaugnet: a new forecasting framework for stock market index value with an overfitting prevention LSTM module and a prediction LSTM module. *Expert Systems with Applications*, 113: 457–480, 2018.

12 Y. Zhang, Q. Yang, Z. Liu, H. Peng, and J. Wang. A prediction model based on gated nonlinear spiking neural system. *International Journal of Neural Systems*, 33(6): 2350029, 2023.

13 S. Fu, Y. Li, S. Sun, and H. Li. Evolutionary support vector machine for RMB exchange rate forecasting. *Physica A*, 521: 692–704, 2019.

14 Y. Wu and J. Gao. Application of support vector neural network with variational mode decomposition for exchange rate forecasting. *Soft Computing*, 23: 6995–7004, 2019.

15 Y. Huang, Y. Gao, Y. Gan, and M. Ye. A new financial data forecasting model using genetic algorithm and long short-term memory network. *Neurocomputing*, 425: 207–218, 2021.

16 G. Wang, T. Tao, and J. Ma. An improved ensemble learning method for exchange rate forecasting based on complementary effect of shallow and deep features. *Expert Systems with Applications*, 184: 115569, 2021.

17 Y. Yang, J. Guo, S. Sun, and Y. Li. Forecasting crude oil price with a new hybrid approach and multi-source data. *Engineering Applications of Artificial Intelligence*, 101: 104217, 2021.

18 L. Yu, S. Wang, and K.K. Lai. Forecasting crude oil price with an EMD-based neural network ensemble learning paradigm. *Energy Economics*, 30: 2623–2635, 2008b.

19 S. Urolagin, N. Sharma, and T.K. Datta. A combined architecture of multi-variate LSTM with Mahalanobis and Z-score transformations for oil price forecasting. *Energy*, 231: 120963, 2021.

20 Y. Zhao, J. Li, and L. Yu. A deep learning ensemble approach for crude oil price forecasting. *Energy Economics*, 66: 9–16, 2017.

21 L. Yu, X. Zhang, and S. Wang. Assessing potentiality of support vector machine method in crude oil price forecasting. *Eurasia Journal of Mathematics Science and Technology Education*, 13(12): 7893–7904, 2017.

22 R. Jammazi and C. Aloui. Crude oil price forecasting: experimental evidence from wavelet decomposition and neural network modeling. *Energy Economics*, 34(3): 828–841, 2012.

23 R. Li, Y. Hu, J. Heng, and X. Chen. A novel multiscale forecasting model for crude oil price time series. *Technological Forecasting and Social Change*, 173: 121181, 2021.

24 L.T. Zhao, S.G. Wang, and Z.G. Zhang. Oil price forecasting using a time-varying approach. *Energies*, 13(6): 1403, 2020.

25 Y. Lin, K. Chen, X. Zhang, B. Tan, and Q. Lu. Forecasting crude oil futures prices using BiLSTM-Attention-CNN model with wavelet transform. *Applied Soft Computing*, 130: 109723, 2022.

7

Power Load Forecasting and Photovoltaic Power Forecasting

7.1 Power Load Forecasting

7.1.1 Introduction

Accurate prediction of electricity loads can significantly improve the economic dispatch and energy resource allocation efficiency, and also enhance the level of power system planning. According to dierent time scales, load forecasting is generally categorized into following types: medium-long-term load forecasting (MLTLF), short-term load forecasting, and very-short-term load forecasting. MLTLF usually predicts future load from few weeks, months, and one to several years, so is suitable for production equipment maintenance, production and operation planning, power scheduling, developing long-term operation plans, expansion and renovation, etc. With the change of urban population, urban economic development, natural environment, and international energy market, etc., the power grid is more complex, making the prediction accuracy affected by uncertainty and volatility. To enhance the prediction accuracy and reliability, researchers have developed different medium-long-term forecasting methods, mainly including: statistically-based approaches, machine learning approaches, deep learning approaches, physical model-based approaches and hybrid models. Conventional statistically-based methods rely on mathematical analysis of historical data, focusing on the model's stability and explanatory power, such as time series analysis method, Kalman filtering method, support vector regression (SVR), linear regression or nonparametric regression method, and trend extrapolation method. However, since electricity load is influenced by numerous factors, relying solely on time series models may not yield optimal results. Machine learning

Spiking Neural P Systems for Time Series Analysis, First Edition. Jun Wang and Hong Peng.

methods include support vector regression [1], fuzzy analysis [2], artificial neural networks [3], and deep neural networks [4], etc. Common deep learning techniques consist of convolutional neural networks (CNN) [5], deep belief networks (DBN) [6], recurrent neural networks (RNN) [7], gated recurrent units (GRU) [5], and long short-term memory networks (LSTM) [8]. These deep learning methods excel in data processing, feature extraction and generalization, often providing superior predictive results compared to traditional methods [9]. Nonetheless, these forecasting techniques also have limitations. For instance, CNNs typically require large datasets for training to maximize their feature learning capabilities, which can be challenging and expensive for tasks with limited samples. Additionally, training deep RNNs can encounter issues with gradient propagation during backpropagation, particularly with long sequences, leading to problems like gradient vanishing or exploding. GRUs may also struggle with information retention, potentially losing important contextual details.

Effective load forecasting fundamentally requires computational models to accurately interpret time-dependent variations and intricate, nonlinear features embedded in power load records. As the nonlinear extension of spiking neural P (SNP) systems [10], nonlinear spiking neural P (NSNP) systems [11] are used to deal with time series prediction problems. What makes NSNP architectures exceptional is their nonlinear spike generation capability, differentiating them from standard computing paradigms. Within the NSNP framework, every neural unit maintains an intrinsic status governed by nonlinear activation principles, where the status gets updated through spiking mechanisms. Therefore, the NSNP systems are suitable for modeling dynamical systems. Combining the NSNP model with attention mechanism, this section introduces a load forecasting model using an improved NSNP system to improve the prediction accuracy of the model [12].

7.1.2 Model

This section provides a comprehensive explanation of the proposed predictive modeling structure, and its overall framework diagram is shown in Figure 7.1, which consists of the following main parts.

7.1.2.1 Data Collection and Preprocessing

First, the collected raw data are interpolated to fill in the missing values. Second, the climate data are resampled to ensure that the load data matches the data granularity of the climate characteristics. Finally, the data are feature coded.

The collected data includes historical load data and historical climate data. The dates corresponding to the load data are labeled, with Monday, Tuesday, ..., and Sunday assigned the values 1,2, ..., and 7, respectively. Weekdays and holidays

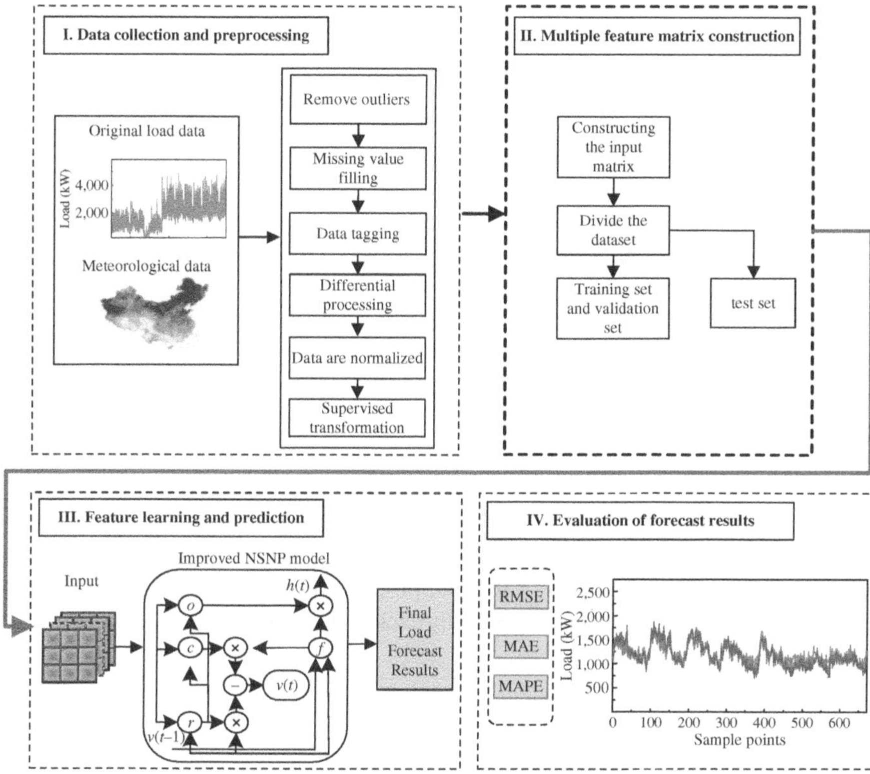

Figure 7.1 Framework diagram of the prediction methodology. Adapted from [12].

are also labeled, with weekdays marked as 0 and official holidays and weekends marked as 1. The temperature data corresponding to the load at each time point are labeled, where Load_t represents the load at time t and Temp_t represents the temperature at time t. If time t falls on a Wednesday, Week_t is set to 3. If time t falls on a weekend or an official holiday, Work_t is set to 1. Figure 7.2 shows data tagging. The dataset consists of one column of load data, one column of temperature data, one column of weekly labels, and one column of holiday labels.

First, the original load data are cleaned, and missing data points are replaced by applying linear interpolation techniques. Subsequently, first-order differencing is applied to the load data to reduce trends and scale effects, resulting in a more stationary series that can be more effectively learned by the model. To further accelerate model convergence during training, the data are normalized to [0, 1], and the scaler used for normalization is saved for later application to inverse transform the model outputs.

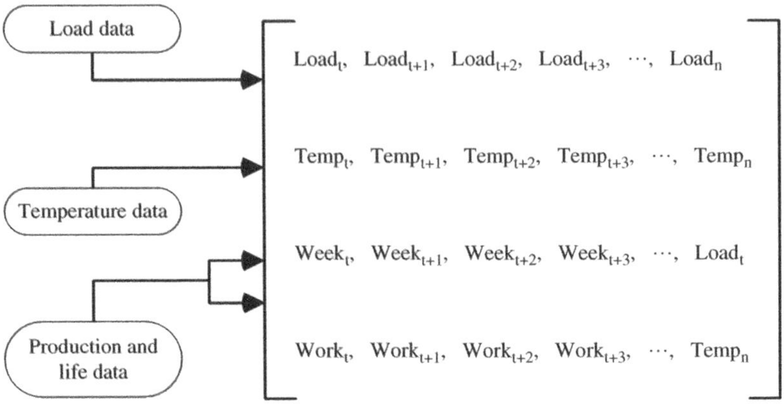

Figure 7.2 Data tagging. Adapted from [12].

7.1.2.2 Multiple Feature Matrix Construction

The single-column load data is transformed into two-column lagged supervised data, enabling the model to learn the mapping relationships across different time steps. The dataset is divided into mutually exclusive training and testing sets based on the chronological order.

7.1.2.3 Predictions Based on NSNP Model

A. Improved NSNP Model

To enhance the model's capacity for extracting temporal patterns and nonlinear relationships from power load datasets, this section develops a NSNP-based forecasting framework (Figure 7.3), building upon the previously introduced NSNP architecture. Moreover, the attention layer is added after the NSNP model.

When dealing with large-scale vision data, people selectively focus their attention on areas that require special attention. The attention mechanism imitates the function of human vision attention, allowing the model to focus on the most important feature part of the input data. Thus, the model reduces or even ignores attention to other unimportant regions, obtaining more important and effective information. The time series data of power load contains many characteristic information with significant patterns. Therefore, the attention mechanism is designed to obtain key information and optimize the efficiency of information processing.

The parameters are defined as follows. The output vector is denoted by $h = [h_1, h_2, \ldots, h_n]$, where n denotes the total number of steps, and h_i denotes the output of step i; the input vector is denoted by $x = [x_1, \ldots, x_n]$, and x_i is the output of step i; the bias vector is denoted by $b = [b_1, b_2, \ldots, b_n]$, where b_i is the bias for step i; the weight vector is denoted by $W = [w_1, \ldots, w_n]$, wherein w_i is the weight

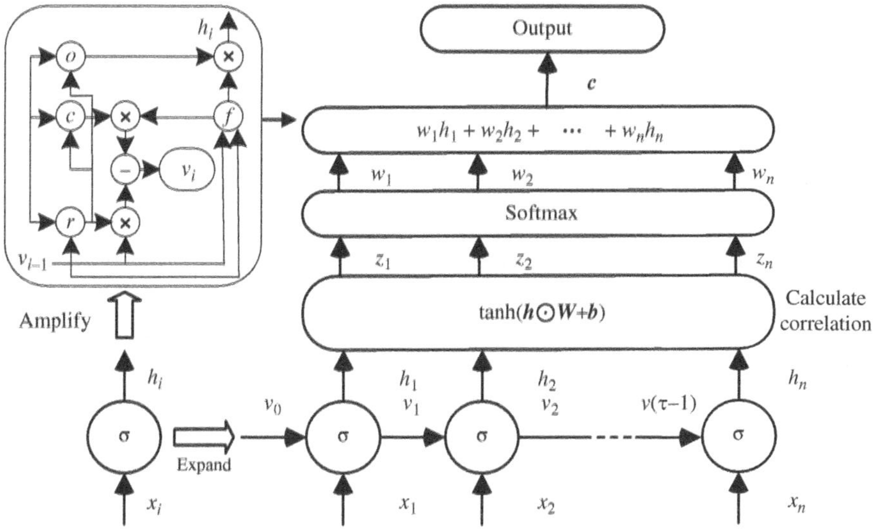

Figure 7.3 The structure of improved NSNP model. Adapted from [12].

of step i; the attention vector is denoted by $\mathbf{z} = [z_1, z_2, \ldots, z_n]$, where z_i is the bias for step i. The trainable parameters are W and b. Within the attention architecture, the layer's input x_i corresponds to the preceding layer's output h_{i-1}.

The output h of the NSNP layer is adopted to calculate the correlation between the outputs at different times, thereby assigning weights to the outputs of the NSNP layer at distinct times. $\mathbf{z}$ is calculated by

$$\mathbf{z} = \tanh\left(h \odot W + b\right) \tag{7.1}$$

By computing the correlation degree between inputs under asynchronous conditions, the weights W are assigned, which represent the correlation between inputs at distinct times. Subsequently, the attention component is normalized using the softmax function to compute the weight w_i at step i, i.e.:

$$w_i = \frac{\exp(z_i)}{\sum_{j=1}^{n} \exp(z_j)} \tag{7.2}$$

And these attention weights are weighted with the output of the previous layer to obtain the output of the model, which is the context vector c. The context vector c is obtained by

$$c = \sum_{j=1}^{n} w_j x_j \tag{7.3}$$

B. Model Training

The model is trained using the training set. After setting the number of epochs, the mean squared error (MSE) of the predicted results at each time step is calculated during each iteration, as below:

$$\text{MSE} = \frac{1}{T}\sum_{i=1}^{T}(y_i - \hat{y}_i)^2 \tag{7.4}$$

wherein y_i and $\hat{y}_i$ represent the true value and predicted value, respectively; T is the length of the prediction sequence.

The MSE value of the predicted results at all time steps is calculated using Eq. (7.4), and the average MSE across all time steps within the epoch is adopted as the loss function. Training is completed when the loss stabilizes and the number of iterations reaches or exceeds the set number of epochs. If either condition is not satisfied, training continues, with the internal model parameters adjusted to minimize the loss until the conditions are met. Upon completion of training, the model is saved for subsequent prediction experiments.

C. Prediction Output

During the testing phase, the well-trained model is used to complete prediction experiments. The model sequentially outputs the prediction results for each time step in the testing set. However, the results are the predicted values after differencing and normalization. Therefore, it is necessary to apply the saved scaler to perform inverse normalization, followed by inverse differencing, in order to obtain the final load prediction values.

7.1.3 Simulation Experiments and Analysis

To evaluate the predictive performance and effectiveness of the parameterized NSNP forecasting model, experiments are conducted using the ISO load dataset provided by the Independent System Operator of New England (ISO-NE) in the United States. The ISO dataset includes regions such as Western/Central Massachusetts (WCMASS), Southeast Massachusetts (SEMASS), Northeast Massachusetts (NEMASS), Vermont (VT), Rhode Island (RI), New Hampshire (NH), Maine (ME), and Connecticut (CT). The experimental parameter settings are shown in Table 7.1.

Three sets of comparative experiments are designed based on a multivariate forecasting model using the ISO dataset. Apart from historical load data, the training data also includes temperature, weekday labels, and workday and holiday labels.

In order to comprehensively assess the prediction performance of the discussed NSNP model, three evaluation metrics are used: absolute percentage error

Table 7.1 Experimental parameter setting. Adapted from [12].

Parameter name	Value
activation	tanh
recurrent_activation	hard_sigmoid
kernel_initializer	glorot_uniform
recurrent_initializer	orthogonal
bias_initializer	zeros
number of neurons	8
Epoch	100
loss	MSE
optimizer	adam
output Layer	Dense(1)

(MAPE), mean absolute error (MAE), and root mean square error (RMSE). MAPE is the mean absolute percentage error, reflecting the average deviation between the actual and predicted values [13]. The MAPE is calculated by

$$\mathrm{MAPE} = \frac{1}{T}\sum_{i=1}^{T}\left|\frac{\hat{y}_i - y_i}{y_i}\right| \tag{7.5}$$

Usually, a smaller MAPE value means better model performance.

The MAE, also known as the L_1-norm loss, is the average of the absolute values of the prediction errors, representing the average difference between the actual and predicted values [13]. The MAE is calculated by

$$\mathrm{MAE} = \frac{1}{T}\sum_{i=1}^{T}|\hat{y}_i - y_i| \tag{7.6}$$

A smaller MAE value indicates more accurate predictions.

The RMSE is the square root of the mean squared prediction error, emphasizing sensitivity to large errors [13]. A smaller RMSE value indicates more accurate model predictions. The RMSE is given by

$$\mathrm{RMSE} = \sqrt{\frac{1}{T}\sum_{i=1}^{T}(\hat{y}_i - y_i)^2} \tag{7.7}$$

In Eqs. (7.5)–(7.7), y_i signifies the actual value, and $\hat{y}_i$ represents the forecasted value.

7.1.3.1 Comparative Experiment I

The original dataset consists of load data, weekday labels, workday or holiday labels, temperature, and other historical data from 2004 to 2011 provided by the ISO system. Three experimental sets are designed: using data from 2009 as the training set and January 2010 as the test set; using data from July 1, 2009, to June 30, 2010, as the training set and July 2010 as the test set; using data from March 2003 to December 2005 as the training set and 2010 to 2011 as the test set.

Figure 7.4 shows the predicted curve for January 2010, and Figure 7.5 shows the predicted curve for July 2010. In the two-month prediction experiment, the predicted curves closely fit the actual load curves, and the distribution of absolute errors remains stable within a certain range, demonstrating the stable predictive performance of the NSNP model. The prediction results for January and July 2010 are shown in Table 7.2. The results indicate that, in the two-month prediction, the NSNP model outperforms two benchmark models in Refs. [14, 15]. Figures 7.6 and 7.7 show the predicted curves for 2010 and 2011, respectively. The predicted results still fit the actual load well, and the error distribution remains stable at a

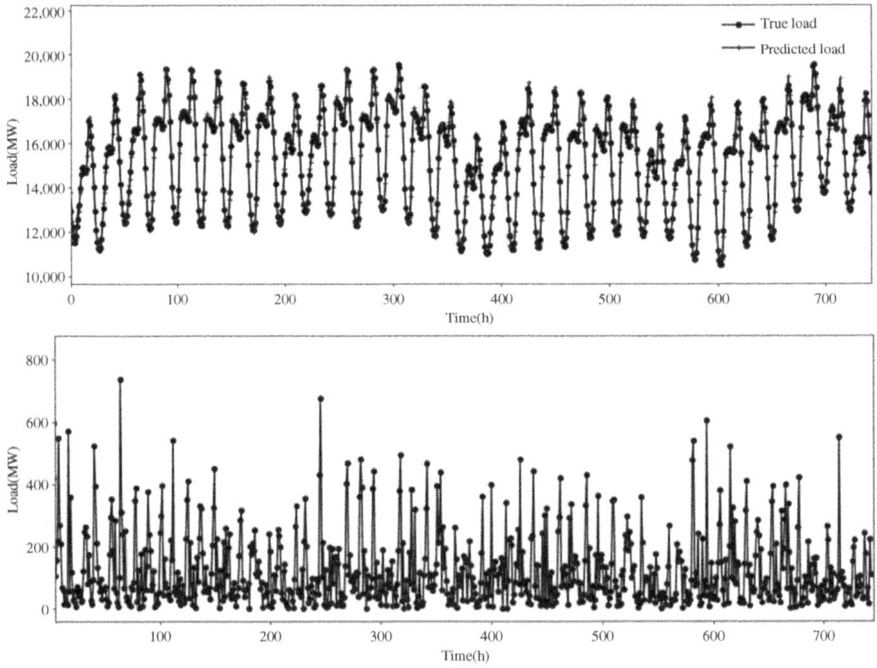

Figure 7.4 Error curves (bottom) and curve fitting (top) in January 2010 for ISO. Adapted from [12].

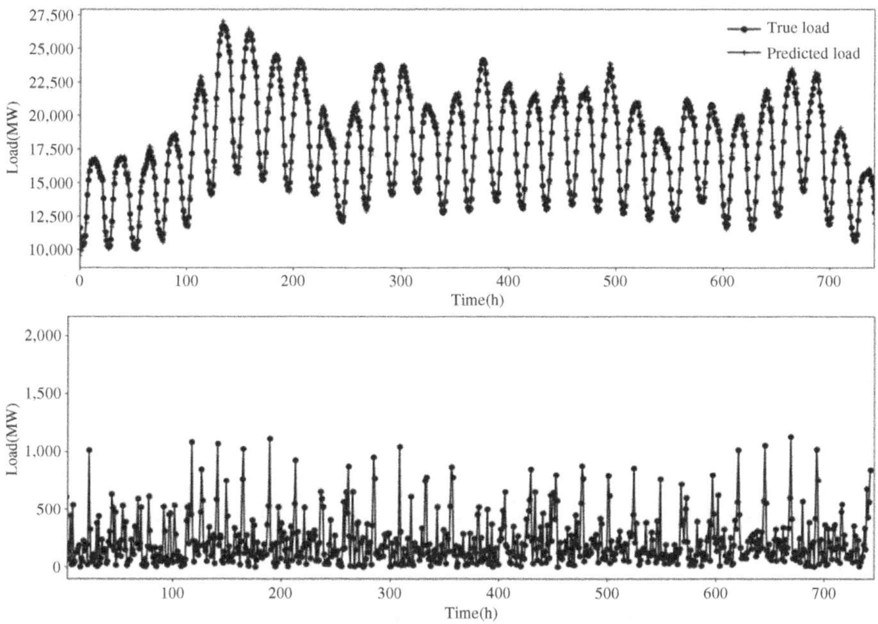

Figure 7.5 Error curves (bottom) and curve fitting (top) in July 2010 for ISO. Adapted from [12].

Table 7.2 Load forecast results for the ISO System for January 2010 and July 2010. Adapted from [12].

	2010.1			2010.7		
Models	**MAPE%**	**MAE**	**RMSE**	**MAPE%**	**MAE**	**RMSE**
ELM-1	1.428	220	303	2.8158	504	708
ELM-2	1.108	170	237	2.3703	433	528
NSNP (our)	0.796	122.384	173.236	1.268	218.621	309.397

low level. The main maxima of the error distribution are consistent with the previous experimental results, with most errors occurring near the peaks and valleys of the curves.

Table 7.3 shows the prediction results on MAPE metric for 2010–2011. In the 2011 predictions, the NSNP achieves the best ranking, while in 2010, it ranks second. This experiment is compared with the methods in Refs. [14, 15], and it can be observed that the NSNP model's prediction performance is stable and accurate, outperforming other forecasting methods.

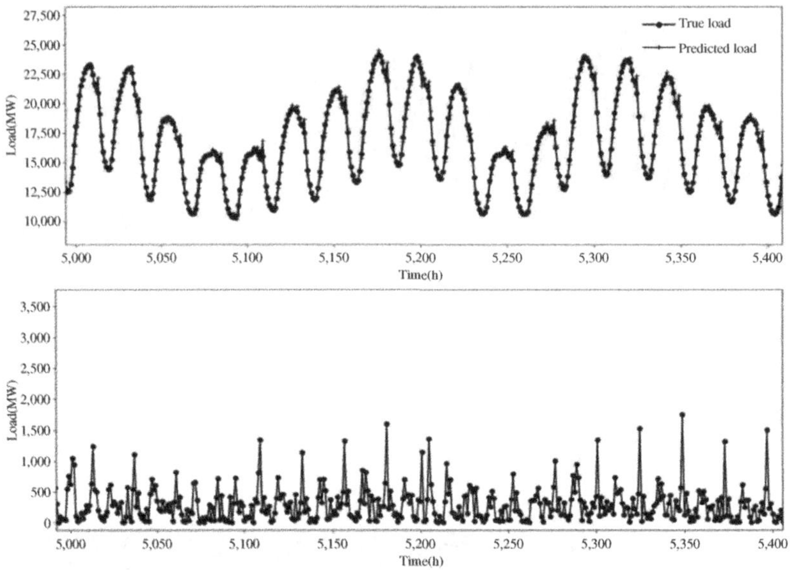

Figure 7.6 Error curves (bottom) and curve fitting (top) in 2010 for ISO. Adapted from [12].

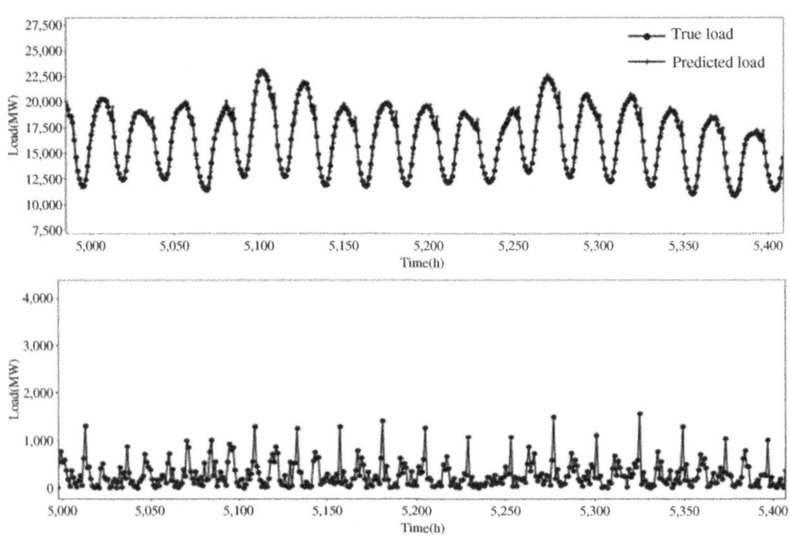

Figure 7.7 Error curves (bottom) and curve fitting (top) in 2011 for ISO. Adapted from [12].

Table 7.3 Load forecast results on MAPE% for 2010 and 2011. Adapted from [12].

Models	2010	2011
SVR (24H)	1.79	2.07
ELM (24H)	1.83	2.19
ISO (24H)	1.95	2.20
ErrCor original (24H)	1.80	2.02
ErrCor modified (24H)	1.75	1.98
ELM-1 (24H)	1.50	1.80
NSNP (our)	1.602	1.586

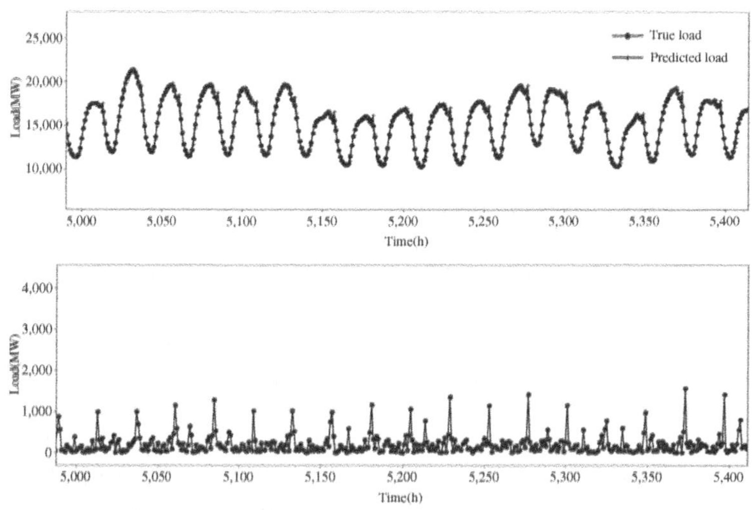

Figure 7.8 Error curves (bottom) and curve fitting (top) for 2013 load forecast results for ISO systems. Adapted from [12].

7.1.3.2 Comparative Experiment II

The original dataset consists of load and temperature historical data from the ISO system from 2009 to 2013. Data from 2009 to 2012 are adopted as the training set, and 2013 are adopted as the test set. Figure 7.8 shows the predicted curve for 2013. In this experiment, the NSNP maintains a high level of fit with the actual load and keeps the absolute error stable within a low range. Table 7.4 presents the prediction results for 2013. When compared with the 11 forecast combination methods in Ref. [16], the NSNP model shows the best performance in terms of the MAPE metric.

Table 7.4 ISO system load forecast MAPE (%) for 2013. Adapted from [12].

Models	MAPE%
PW	2.12
LAD	2.14
OLS	2.14
WA	2.10
TA	2.10
Simple	2.10
CLS	2.11
IRMSE	2.10
EWA	2.18
FS	2.11
ML-poly	2.11
NSNP	1.543

7.1.3.3 Comparative Experiment III

The original dataset consists of load, weekday labels, workday or holiday labels, and temperature historical data from the ISO system from 2004 to 2009. Data from 2004 to 2008 is adopted as the training set, and 2009 is adopted as the test set.

In the prediction results, the forecasts for five holidays are selected and compared with three forecasting methods from Ref. [17], including ARIMA, BPNN, and ENFF. The selected dates are Christmas (December 25, 2009, Friday), Memorial Day (May 25, 2009, Monday), Easter (April 12, 2009, Sunday), Labor Day (September 7, 2009, Monday), and New Year's Day (January 1, 2009, Thursday). The comparison results are shown in Table 7.5. Figure 7.9 shows some of the predicted curves for 2009. The extreme error values are mostly distributed near the peaks and valleys of the load curve.

In the three sets of multivariate experiments, the prediction results of the discussed NSNP model are compared with those of the benchmark models. The experimental results demonstrate that the discussed NSNP model achieves lower prediction errors than the benchmark models in the comparison results, ranking first and second in all comparisons. In each experimental set, the discussed NSNP model demonstrates the best overall forecasting performance. This indicates that the discussed NSNP model has strong load prediction capabilities.

Table 7.5 Load forecast results on MAPE (%) for 2009. Adapted from [12].

Models	Christmas	Memorial Day	Labor Day	Easter	New Year's Day
ARIMA	5.61	4.51	5.22	2.52	3.27
BPNN	4.57	3.89	3.90	5.22	4.96
ENFF	1.85	2.36	1.87	1.77	2.15
NSNP (our)	1.598	2.293	2.262	2.028	1.472

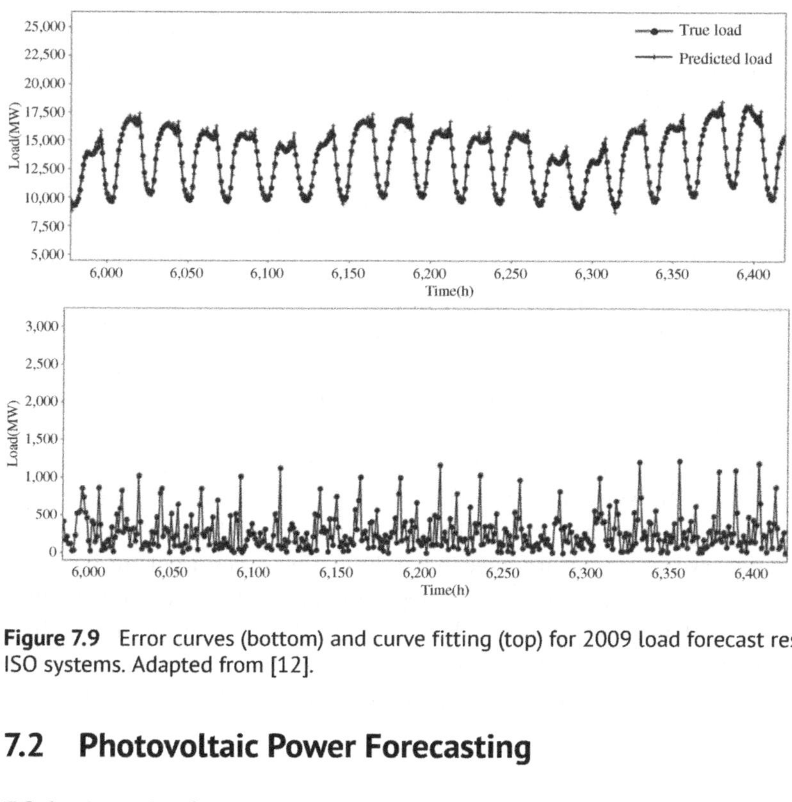

Figure 7.9 Error curves (bottom) and curve fitting (top) for 2009 load forecast results for ISO systems. Adapted from [12].

7.2 Photovoltaic Power Forecasting

7.2.1 Introduction

Driven by technological advancements and cost reductions, the global adoption rate of solar energy is rapidly increasing, making significant contributions to the development of clean energy. As one of the main forms of solar energy, the growth rate of photovoltaic power generation is accelerating every year. However, photovoltaic power generation has inherent instability and intermittency, as it relies on sunlight exposure, which is unstable and greatly influenced by weather, seasons, and time [18]. Various factors, including climate conditions and weather patterns, can significantly affect the output of photovoltaic power generation systems. With the deep integration of smart grid and photovoltaic power generation, these characteristics have brought new challenges to the power system, especially making grid scheduling and management more complex [19]. Therefore, improving the forecasting accuracy of photovoltaic power generation can effectively enhance the efficiency and operational quality of power system scheduling, while minimizing excess capacity to the greatest extent possible.

According to different time scales, photovoltaic power generation prediction is usually divided into the following types: ultra-short term [20], short term, and medium to long term [21, 22]. Short term prediction specifically targets the upcoming fluctuations in photovoltaic power output, typically spanning a time range of minutes to hours. Short term photovoltaic prediction is helpful for the operation and management of smart grids, as the grid adjusts and optimizes in real-time based on the inherent changes in photovoltaic power generation. Moreover, based on different prediction time intervals, photovoltaic power generation prediction technology can also be divided into multi-step forecasting and single step forecasting. In addition, predictions can be classified as regional and single-field predictions, reflecting the spatial scale differences in prediction work [23].

The basic characteristics of photovoltaic power generation data are time series data, which are complex, nonlinear and dynamic. These data exhibit both short-term correlations and long-term dependencies. The inherent nonlinear spiking mechanism in NSNP systems is particularly adopted at handling complex nonlinear data changes. Compared to traditional neural networks, SNP systems and their variants have better performance in handling time series problems by applying nonlinear spiking mechanisms of biological neurons. These variants include GSNP model [24] and ESN model [25]. To improve the accuracy of photovoltaic power generation prediction, this section discusses an ESN variant starting from an improved NSNP system and constructs a forecasting model for short-term PV power generation [26].

7.2.2 Model

7.2.2.1 Structure of Prediction Model

Figure 7.10 shows the framework of the photovoltaic (PV) power generation forecasting model, consisting of three distinct components. The three components are as follows:

1. Data collection and preprocessing: Three datasets downloaded from Australia are preprocessed to preserve the required meteorological data and normalized to an appropriate scale.

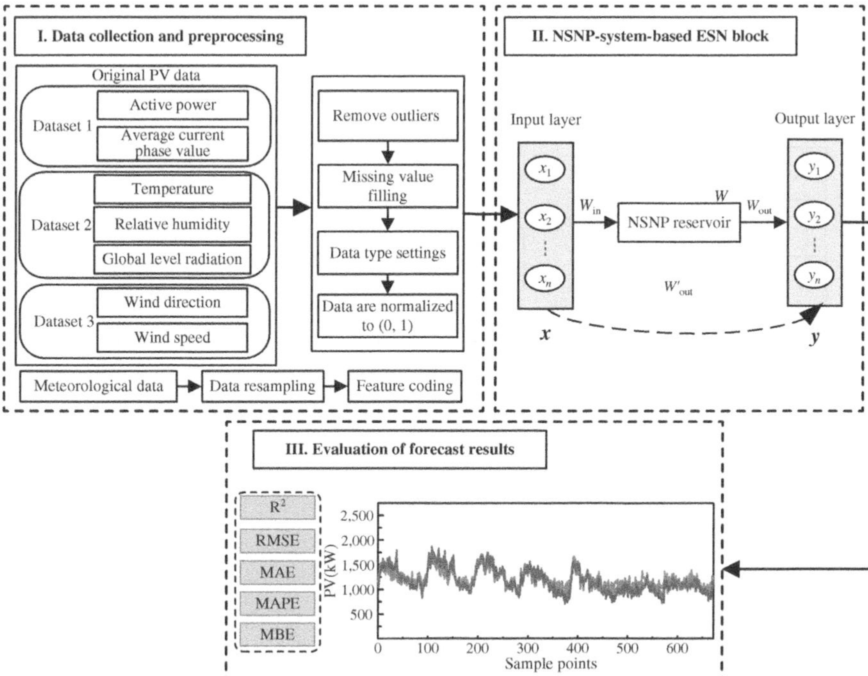

Figure 7.10 The framework of the photovoltaic power generation prediction model. Adapted from [26].

2. NSNP-system-based ESN block: The preprocessed data is input into the constructed photovoltaic power prediction module for training.
3. Evaluation of forecast results: Five evaluation indicators are used to assess the predicted results.

The implementation flow chart of the entire model is shown in Figure 7.11. The working process of this prediction model can be described as follows. First, the data is preprocessed and divided into a training set and a test set. Then the training set is input into the initialized ESN-NSNP model for training, and the state of the NSNP reservoir is updated to optimize the output matrices W_{out} and W'_{out}. Finally, the test phase is entered. For each time step of the test set sequence, its state is updated and the trained W_{out} and W'_{out} are used to obtain the corresponding prediction output. The prediction result at this time is the normalized prediction value, so it is inversely normalized to obtain the final photovoltaic power deterministic prediction output.

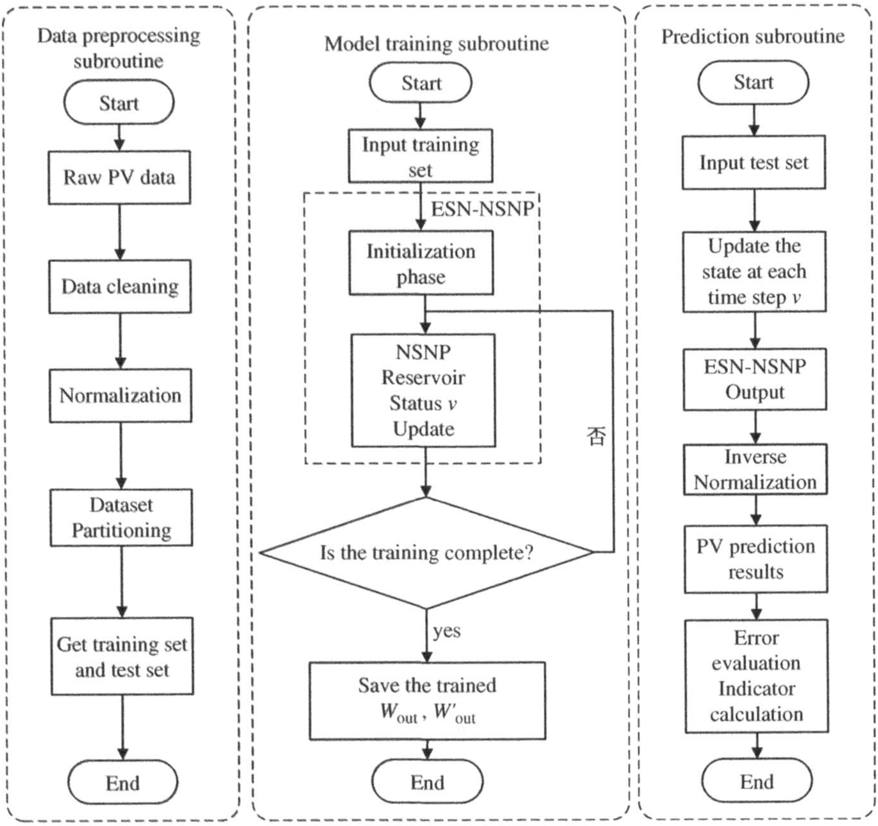

Figure 7.11 The flow chart of the prediction model. Adapted from [26].

7.2.2.2 Detailed Description of ESN–NSNP Block

Due to nonlinear spiking mechanism, NSNP systems can provide strong nonlinear dynamics. This section aims to present a new photovoltaic power forecasting method utilizing the NSNP systems. ESN is a specific subclass of recurrent neural networks (RNNs) that processes sequential input data by combining echo-like neurons in a reservoir framework. This architecture enhances the performance and trainability of ESN in processing sequential data, effectively addressing the limitations of traditional RNN, such as gradient vanishing and training complexity. Structurally, an ESN comprises a reservoir, an input layer and an output (readout) layer, with the reservoir containing a significant number of interconnected neurons. Consequently, the model in this section represents an innovative recurrent-like architecture that integrates an NSNP-system-based ESN block.

Unlike conventional ESNs and their variants, this model utilizes a NSNP system as its reservoir, thereby endowing the framework with richer dynamic characteristics. This NSNP system can be expressed by a directed graph having n neurons, wherein neurons act as nodes and edges represent synaptic connections between them.

Figure 7.12 illustrates a NSNP system, comprising n spiking neurons denoted by $\sigma_1, \sigma_2, \ldots, \sigma_n$. Each neuron has a state unit v_i, an external input x, and a nonlinear rule of the form $\alpha^{g(v)} \rightarrow \alpha^{f(v)}$, where $g(v)$ and $f(v)$ are the spike consumption and spike generation functions, respectively.

Suppose that the state of neuron σ_i at time τ is denoted by $v_i(\tau)$, and its external input is denoted by $x(\tau)$. Upon the firing of neuron σ_i, the system engages in the process of consuming the spikes represented by $g(v)$ and generating the spikes denoted by $f(v)$. Consequently, the state of the neuron is updated based on the equation $v = v_i(\tau - 1) + x(\tau)$. The subsequent neurons will receive the generated spikes. Thus, the state equation for σ_i is given as follows:

$$v_i(\tau) = v_i(\tau - 1) - g(x(\tau) + v_i(\tau - 1)) + \sum_{j=1, j \neq i}^{n} f(x(\tau) + v_j(\tau - 1)) \tag{7.8}$$

wherein $i = 1, \ldots, n$. Assume that the external input is expressed by $x = \sum_{l=1}^{m} x_l$. Consequently, the state equation is given by

$$v_i(\tau) = v_i(\tau - 1) - g\left(\sum_{l=1}^{m} x_l(\tau) + v_i(\tau - 1)\right) + \sum_{j=1, j \neq i}^{n} f\left(\sum_{l=1}^{m} x_l(\tau) + v_j(\tau - 1)\right) \tag{7.9}$$

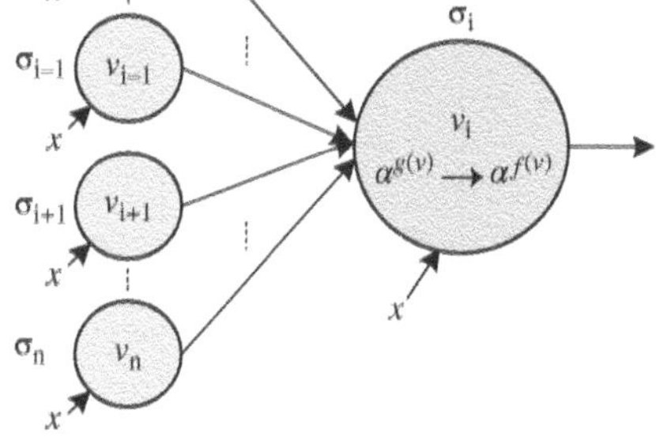

Figure 7.12 A NSNP system [26]/MDPI/CC BY 4.0.

To simplify the model, let $g(v) = -f(v)$. Therefore, the state equation for σ_i is reformulated by

$$v_i(\tau) = v_i(\tau - 1) + \sum_{j=1}^{n} f\left(\sum_{l=1}^{m} x_l(\tau) + v_j(\tau - 1)\right) \tag{7.10}$$

After the parameterizing, one can derive a parameterized NSNP system. The state equation for σ_i is given by

$$v_i(\tau) = \eta \cdot v_i(\tau - 1) + \sum_{j=1}^{n} w_{ij}^1 f\left(\sum_{l=1}^{m} w_{lj}^2 x_l(\tau) + \gamma \cdot v_j(\tau - 1)\right) \tag{7.11}$$

where η, γ, w_{ij}^1, and w_{ij}^2 are defined as parameters, $j = 1, 2, \ldots, m$, and $i = 1, 2, \ldots, n$. Moreover, this state equation can be reformulated in matrix notation as follows:

$$\mathbf{v}(\tau) = \eta \cdot \mathbf{v}(\tau - 1) + Wf(W_{\text{in}}\mathbf{x}(\tau) + \gamma \cdot \mathbf{v}(\tau - 1)) \tag{7.12}$$

wherein $\mathbf{v}(\tau) = (v_1(\tau), \ldots, v_n(\tau))^T$ and $\mathbf{x}(\tau) = (x_1(\tau), \ldots, x_m(\tau))^T$. The parameters η and γ are scalar values, and $W = (w_{ij}^1)_{n\times n}$ and $W_{\text{in}} = (w_{lj}^2)_{m\times n}$ are parameter matrices.

The parameterized NSNP system under consideration is classified as a dynamical system, with Eq. (7.12) serving as its mathematical representation. The system responsible for producing the photovoltaic power data is fundamentally an intricate and unknown dynamical system. The parameterized NSNP system will be employed to model this unidentified complex dynamical system.

Figure 7.13 shows the architecture of the ESN block based on the NSNP system. This block consists of a NSNP system, an input layer and an output (readout) layer.

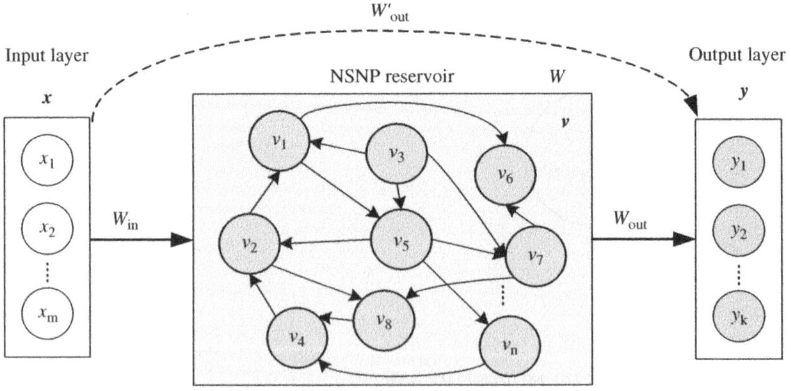

Figure 7.13 ESN-NSNP block [26]/MDPI/CC BY 4.0.

Let us consider a system comprising m input neurons within the input layer, where the inputs at time τ are denoted by the vector $\mathbf{x}(\tau) = [x_1, \ldots, x_m]^T$. The reservoir, characterized by a NSNP system, consists of n spiking neurons. Additionally, the output layer is constituted of k output neurons.

Let us denote $W = (w_{ij}^2)_{n\times n} \in \mathbb{R}^{n\times n}$ as the weight matrix of the reservoir. $W_{\text{in}} = (w_{ij}^1)_{m\times n} \in \mathbb{R}^{m\times n}$ serves as the input-reservoir weight matrix. Additionally, $W_{\text{out}} = (w_{ij}^3)_{n\times k} \in \mathbb{R}^{n\times k}$ is the reservoir-output weight matrix. Finally, $W'_{\text{out}} = (w_{ij}^4)_{m\times k} \in \mathbb{R}^{m\times k}$ is the input-output weight matrix.

The state equation of the discussed model can be expressed as follows:

$$\mathbf{v}(\tau) = \eta \cdot \mathbf{v}(\tau - 1) + Wf(\gamma \cdot \mathbf{v}(\tau - 1) + W_{\text{in}}\mathbf{x}(\tau)) \tag{7.13}$$

and the output equation is given by

$$\mathbf{y}(\tau) = f_o\left(W'_{\text{out}}\mathbf{x}(\tau) + W_{\text{out}}\mathbf{v}(\tau)\right) \tag{7.14}$$

In this context, the function f is characterized by a *sigmoid* function, such as *tanh*, while f_0 is defined as either a nonlinear or linear function; $\eta \in [0, 1]$ and $\gamma \in [0, 1]$ are two scalar parameters; $\mathbf{y}(\tau) = [y_1(\tau), y_2(\tau), \ldots, y_k(\tau)]^T$ denotes the output vector at time τ. Additionally, the weight matrices are specified as $W_{\text{in}} \in \mathbb{R}^{m\times n}$, $W \in R^{n\times n}$, $W_{\text{out}} \in \mathbb{R}^{n\times k}$ and $W'_{\text{out}} \in \mathbb{R}^{m\times k}$. For simplicity, f_0 is given as the identity function, specifically $f_o(x) = x$.

7.2.2.3 Implementation of The Prediction Model

The discussed model employs a supervised learning algorithm for its training process, and it consists of two distinct stages: the initialization stage and the training stage. During the initialization stage, matrices W and W_{in} are generated randomly and remain fixed. The subsequent phase, known as the training stage, aims to optimize the output matrices W_{out} and W_{out} through the training process.

Let $X(\tau) = [\mathbf{v}(\tau)^T, \mathbf{x}(\tau)^T]^T \in \mathbb{R}^{n+m}$, and $\overline{W}_{\text{out}}(\tau) = [W_{\text{out}}^T, W_{\text{out}}'^T]^T$. $\mathbf{x}(\tau) = [x_1(\tau), \ldots, x_m(\tau)]^T \in \mathbb{R}^m$ is the input vector at time τ, while $\mathbf{y}(\tau) = [y_1(\tau), \ldots, y_k(\tau)]^T \in \mathbb{R}^k$ signifies the corresponding output. Consequently, the output of the discussed model is given as follows:

$$\mathbf{y}(\tau) = \overline{W}_{\text{out}}(\tau)X(\tau) \tag{7.15}$$

Let us denote the desired output sequences by $\mathbf{y}(1), \ldots, \mathbf{y}(M)$, and assume that there exist M corresponding sequence data points represented by $\mathbf{x}(1), \ldots, \mathbf{x}(M)$. The matrix H is defined as $H = [\mathbf{x}(1), \ldots, \mathbf{x}(M)]$. Similarly, the matrix $Y = [\mathbf{y}(1), \ldots, \mathbf{y}(M)]$ is defined. Consequently, the relationship between these matrices can be expressed as follows:

$$Y = \overline{W}_{\text{out}}H \tag{7.16}$$

where $\overline{W}_{\text{out}}$ is determined by the equation:

$$\overline{W}_{\text{out}} = (H^T H)^{-1} HY \tag{7.17}$$

where $(H^T H)^{-1}$ stands for the inverse matrix of $H^T H$. It is important to note that a high dimensionality of the matrix H can lead to model overfitting. To mitigate this issue, a regularization term can be incorporated. Thus, a regularized objective function is given as follows:

$$J(W_{\text{out}}) = \frac{1}{2} \| Y - \overline{W}_{\text{out}} H \|^2 + \frac{1}{2} \lambda \| \overline{W}_{\text{out}} \|^2 \tag{7.18}$$

wherein the variable λ is the regularization factor. Consequently, $\overline{W}_{out}$ can be expressed by

$$\overline{W}_{\text{out}} = (H^T H + \lambda I)^{-1} HY \tag{7.19}$$

wherein $I \in R^{m \times m}$ is the identity matrix. Within the framework of the discussed model, which bears resemblance to an ESN, the ridge regression algorithm is employed to determine the $\overline{W}_{\text{out}}$ as follows:

$$W_{\text{out}}^{*} = \arg \min_{W_{\text{out}}} \left\{ J(W_{\text{out}}) = \frac{1}{2} \| Y - W_{\text{out}} H \|^2 + \frac{1}{2} \lambda \| W_{\text{out}} \|^2 \right\} \tag{7.20}$$

After obtaining the optimal weight parameters, the proposed model can effectively forecast test dataset values.

7.2.3 Model Evaluation

The proposed model's forecasting accuracy is evaluated using three distinct PV system datasets serving as performance benchmarks. These datasets have been widely applied in related work. In order to facilitate a comparison between the discussed model and baseline prediction models, the experiments on the three datasets are conducted as three cases.

7.2.3.1 Case Study I

Dataset

The dataset utilized in this case study is derived from the DKASC Alice Springs photovoltaic system site 2 [27]. Data collection for this system occurs at 5-minute intervals, resulting in a comprehensive dataset comprising 103,954 samples for the entirety of the year 2019. The five input variables selected for examination include the average current phase, active power, relative humidity, ambient temperature, and global horizontal irradiance. Significantly, both the mean current phase and global horizontal irradiance averages have a straightforward impact on photovoltaic energy output. The original data is divided into training and testing subsets, with 90% assigned for model training and the remaining 10% reserved for evaluation.

Performance Metrics

Three metrics, R^2, RMSE, and MAE, are used to verify the efficacy of the model. The definition of MAE and RMSE can be found in Eqs. (7.6) and (7.7), respectively. The definition of R^2 metric is provided below:

$$R^2 = 1 - \frac{\sum_{i=1}^{N}(y_i - \bar{y}_i)^2}{\sum_{i=1}^{N}(y_i - \bar{y}_{\text{period}})^2} \tag{7.21}$$

wherein y_i signifies the actual value, and $\bar{y}_i$ represents the forecasted value; $\bar{y}_{\text{period}}$ indicates the average value during the prediction interval.

Experimental Results

To comprehensively demonstrate the superiority of the discussed model, the dataset is processed at three distinct resolutions: 5-, 10-, and 15-minute intervals. The discussed model is evaluated against four baseline models, which include CNN [28], CNN_LSTM [29], QK_CNN [30], and QK_CNN (FC) [30].

Table 7.6 presents the optimal hyperparameter values employed in Case I. A grid search methodology is implemented to identify the most effective values for all hyperparameters. The six hyperparameter configurations within the discussed model are delineated as below: (i) λ is the regularization factor; (ii) δ and μ correspond to the proportion of nonzero connections and the spectral radius, respectively; (iii) n signifies the number of neurons; (iv) η and γ denote two scalar factors.

Tables 7.7–7.9 present the comparison of the discussed model with four baseline methods, detailing results at resolutions of 15, 10, and 5 minutes. The experimental data for the baseline methods are sourced from the original literature. At both 15- and 5-minute resolutions, the discussed model demonstrates superior performance across all evaluation metrics compared to the four baseline methods. In the

Table 7.6 The hyperparameters for the discussed model [26]/MDPI/CC BY 4.0.

Resolution (min)	η	γ	n	μ	δ(%)	λ
15	0.31	0.545	51	0.98	9.5	1.0×10^{-6}
10	0.70	0.450	51	0.98	45	1.0×10^{-6}
5	0.75	0.400	51	0.98	9.5	1.0×10^{-6}

Table 7.7 Comparison results of different models at 15-minute resolution [26]/MDPI/CC BY 4.0.

Models	RMSE	MAE	R^2
CNN	0.529	0.254	0.96
CNN_LSTM	0.547	0.268	0.96
QK_CNN(FC)	0.529	0.236	0.96
QK_CNN	0.519	0.230	0.96
Our model	0.492	0.064	0.96

Table 7.8 Comparison results of different models at 10-minute resolution [26]/MDPI/CC BY 4.0.

Models	RMSE	MAE	R^2
CNN	0.449	0.197	0.97
CNN_LSTM	0.453	0.200	0.97
QK_CNN(FC)	0.453	0.181	0.97
QK_CNN	0.448	0.178	0.97
Our model	0.452	0.058	0.97

Table 7.9 Comparison results of different models at 5-minute resolution [26]/MDPI/CC BY 4.0.

Models	RMSE	MAE	R^2
CNN	0.353	0.133	0.98
CNN_LSTM	0.352	0.130	0.98
QK_CNN (FC)	0.370	0.138	0.98
QK_CNN	0.351	0.124	0.98
Our model	0.348	0.039	0.98

case of the 10-minute resolution, the discussed model exceeds the baseline methods in terms of MAE and R^2 metrics; however, it exhibits slightly inferior performance relative to the CNN and QK_CNN models in the RMSE metric.

Figures 7.14–7.16 presents the prediction results of the discussed model on the test set at resolutions of 15, 10, and 5 minutes, respectively. The top side of each panel displays the original and predicted curves for each group of PV power within the test set, while the bottom side illustrates the corresponding absolute error curves. An examination of these figures reveals that the forecasting curves closely

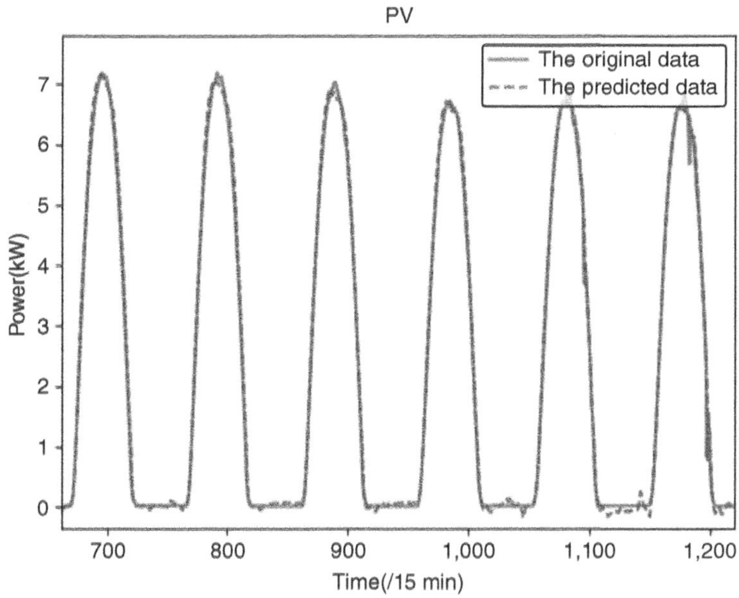

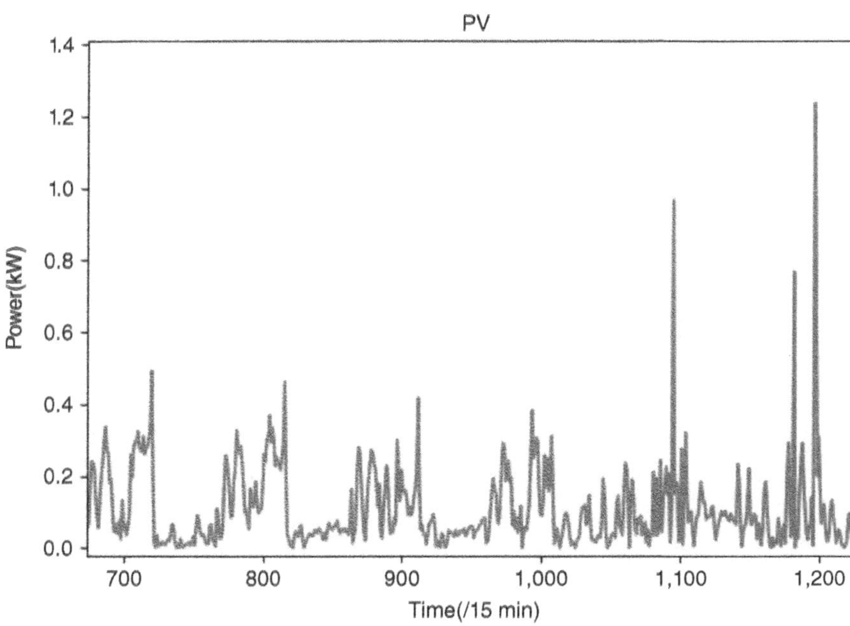

Figure 7.14 Prediction results of the discussed model at 15-minute resolution [26]/MDPI/CC BY 4.0.

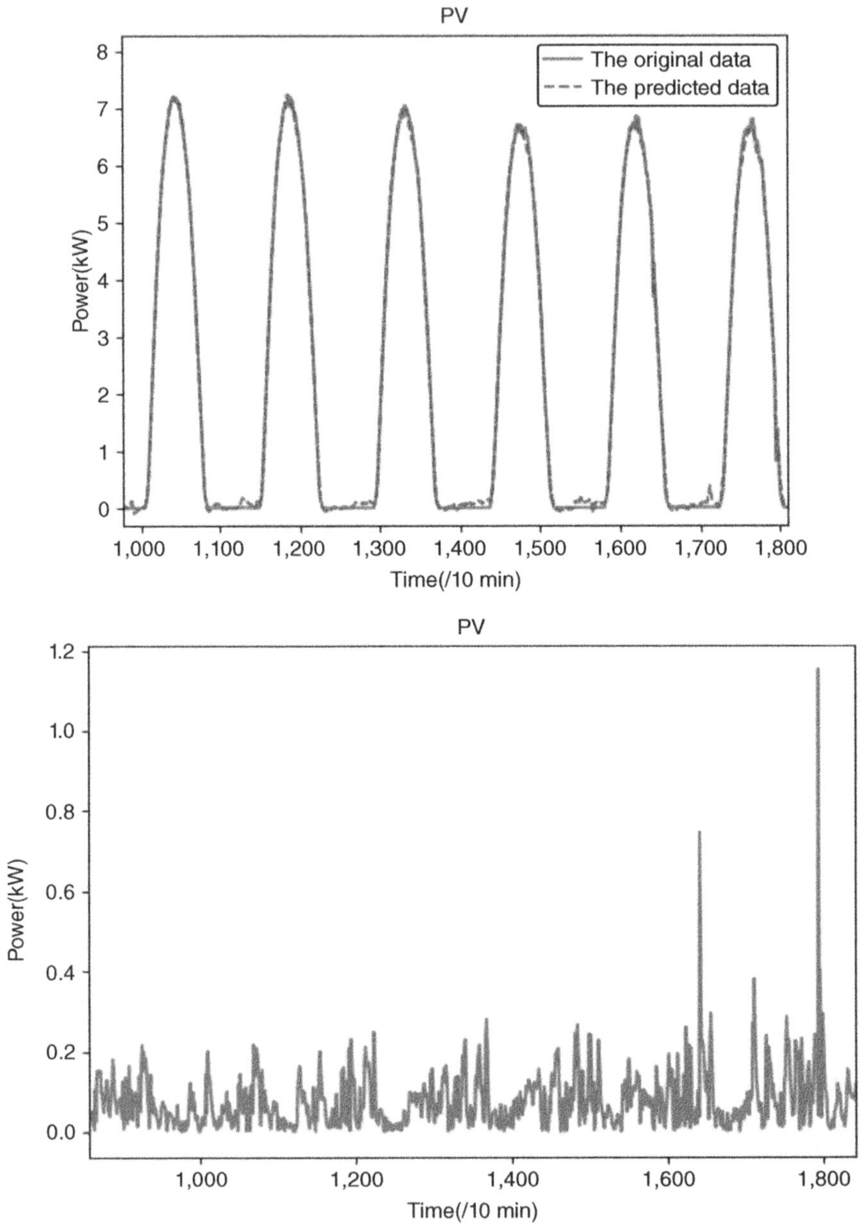

Figure 7.15 Prediction results of the discussed model at 10-minute resolution [26]/MDPI/CC BY 4.0.

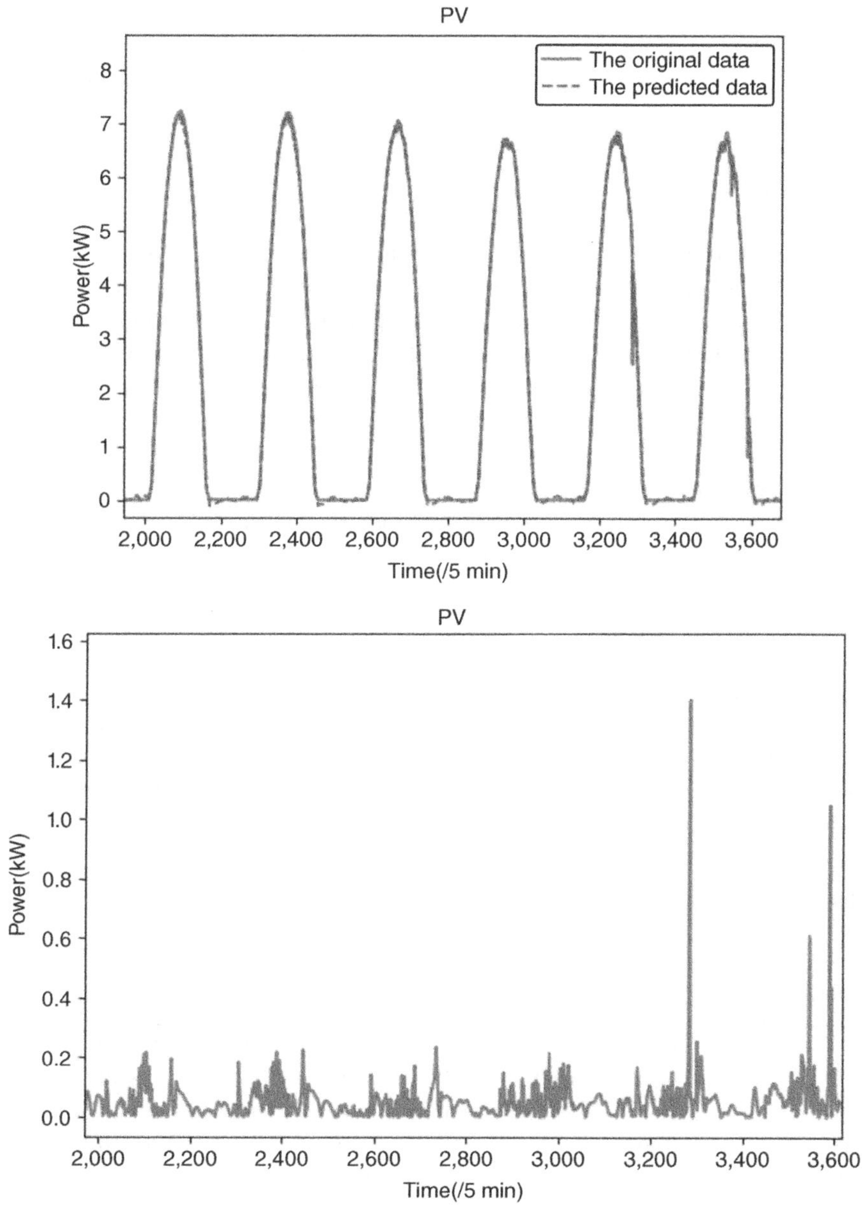

Figure 7.16 Prediction results of the discussed model at 5-minute resolution [26]/MDPI/CC BY 4.0.

align with the original curves across all three resolutions, indicating that the discussed model effectively captures the fluctuations present in the PV data. Furthermore, the absolute prediction errors are notably lower at each of the three resolutions.

7.2.3.2 Case Study II

Dataset

The dataset utilized in this case study is sourced from 4 years of data (2014–2017) collected at Alice Springs 1B in central Australia, with a temporal resolution of 5 minutes [27]. This dataset encompasses 12 distinct attributes, which include current phase average, horizontal diffuse radiation, wind speed, climatic relative humidity, climatic temperature, active power, among others. All available attributes are employed as inputs for the model.

Performance Metrics

In order to verify the performance of the methods, the metrics of MAE, MAPE, and RMSE are employed as assessment criteria for the experiments. The definitions of MAE, MAPE, and RMSE can be found in Eqs. (7.5–7.7), respectively.

Experimental Results

Table 7.10 gives the parameter configuration for the discussed model.

The discussed model is evaluated against three baseline methods (LSTM, CNN, and CLSTM) across eight distinct input sequences, with the results of this comparison presented in Table 7.11. The experimental data for the baseline models can be retrieved in Ref. [29]. The findings from Table 7.11 indicate the

Table 7.10 Parameters configuration of the discussed model Adapted from Guo et al. [26].

Input sequence	η	γ	n	μ	δ(%)	λ
0.5Y	0.20	0.98	15	0.98	40	1.0×10^{-6}
1Y	0.70	0.60	51	0.98	10	1.0×10^{-6}
1.5Y	0.50	0.10	51	0.98	10	1.0×10^{-6}
2Y	0.09	0.50	51	0.98	9	1.0×10^{-6}
2.5Y	0.09	0.60	51	0.98	50	1.0×10^{-6}
3Y	0.05	0.40	51	0.98	35	1.0×10^{-6}
3.5Y	0.90	0.40	51	0.98	35	1.0×10^{-6}
4Y	0.90	0.50	51	0.98	35	1.0×10^{-6}

following: (i) the discussed model achieves the lowest values for both RMSE and MAE across all eight input sequences; (ii) LSTM records the best MAPE value for the 1Y input sequence, CNN excels for the 1.5Y input sequence, and CLSTM attains the best MAPE values for the remaining six input sequences. Overall, while the discussed model exhibits superior forecasting capabilities as evidenced by the RMSE and MAE metrics, its performance in terms of the MAPE metric is comparatively weaker. Furthermore, Table 7.11 illustrates that the accuracy of the LSTM, CNN, and CLSTM models does not consistently improve with increasing input sequence length; rather, it tends to decline. This phenomenon may be attributed to the inherent characteristics of the time series data. As the volume of data processed increases, the temporal distance between the preceding input data and the corresponding output data also increases, resulting in a diminished correlation between them. Consequently, the addition of more data may lead to a deterioration in predictive performance rather than an enhancement. Comparatively, the proposed architecture shows substantially weaker correlation between input size and prediction performance than alternative models. Such stability primarily results from its NSNP component's advanced feature extraction and long-range dependency modeling capacities.

Figures 7.17 and 7.18 illustrate the predictive performance of the discussed model at 0.5 and 4Y, respectively. On the top side of each figure, the original and predicted curves for each group of PV power are displayed, while the bottom side presents the corresponding absolute error curves. These results indicate that the discussed model possesses a strong capacity to accurately capture the fluctuations within the PV data series, demonstrating a minimal absolute prediction error.

7.2.3.3 Case Study III

Dataset

The dataset utilized in this case study is derived from DKASC in Alice Springs [27], Australia, covering the period from June 1, 2014, to June 12, 2016. Six variables are selected as inputs for the discussed model: diffuse horizontal radiation, photovoltaic power, relative humidity, wind speed, global horizontal radiation, and ambient temperature. The training dataset encompasses observations from June 1, 2014, to May 31, 2015, while the testing dataset includes data from June 1, 2015, to June 12, 2016. Consequently, the training set comprises 365 days of weather conditions and PV power generation data, whereas the test set consists of 378 days of similar data.

Table 7.11 Outcomes of predictions generated by various models Adapted from [26].

Input sequence	Metrics	CNN	LSTM	CLSTM	Our model
0.5Y	MAE	0.566	0.654	0.559	0.018
	RMSE	1.128	1.244	1.161	0.084
	MAPE	0.114	0.131	0.112	0.301
1Y	MAE	0.640	0.616	0.628	0.025
	RMSE	1.563	1.393	1.434	0.132
	MAPE	0.111	0.103	0.105	0.474
1.5Y	MAE	0.567	0.599	0.529	0.016
	RMSE	1.411	1.533	1.248	0.111
	MAPE	0.095	0.101	0.095	0.355
2Y	MAE	0.452	0.457	0.397	0.020
	RMSE	0.983	1.320	0.941	0.132
	MAPE	0.059	0.068	0.052	0.662
2.5Y	MAE	0.231	0.389	0.198	0.027
	RMSE	0.447	0.945	0.426	0.132
	MAPE	0.041	0.051	0.035	0.493
3Y	MAE	0.140	0.181	0.126	0.016
	RMSE	0.367	0.398	0.343	0.131
	MAPE	0.025	0.032	0.022	0.519
3.5Y	MAE	0.412	0.455	0.384	0.013
	RMSE	1.136	1.150	0.991	0.083
	MAPE	0.077	0.083	0.070	0.461
4Y	MAE	0.478	0.565	0.405	0.010
	RMSE	0.971	1.465	0.886	0.079
	MAPE	0.083	0.089	0.080	0.707

Performance Metrics

Three evaluation metrics are utilized for the purpose of comparison with the baseline models: RMSE, MAPE, and MBE. The formulations for MAPE and RMSE are presented in Eqs. (7.5) and (7.7), respectively. The MBE metric is defined by

$$\mathrm{MBE} = \frac{1}{N}\sum_{i=1}^{N}(y_i - \bar{y}_i) \tag{7.22}$$

wherein N denotes the forecast horizon; y_i and $\bar{y}_i$ signify the predicted output and the actual value, respectively.

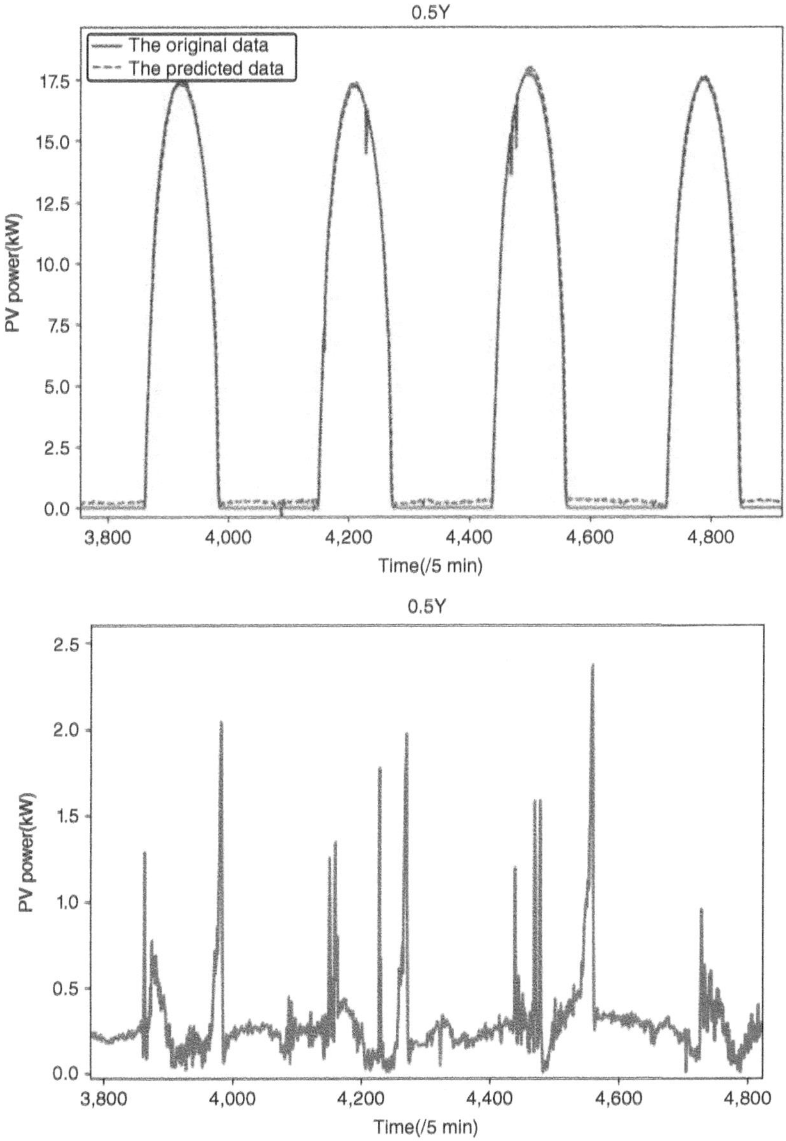

Figure 7.17 Prediction results of the discussed model on 0.5Y [26]/MDPI/CC BY 4.0.

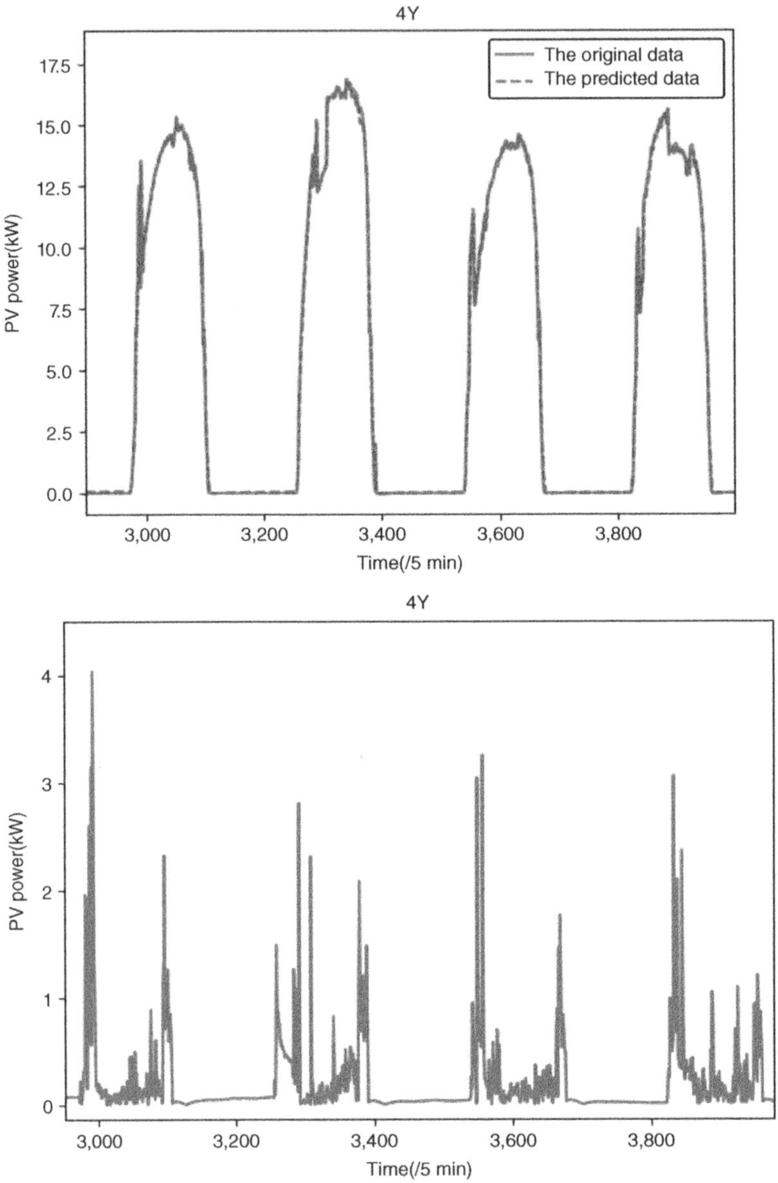

Figure 7.18 Prediction results of the discussed model on 4Y [26]/MDPI/CC BY 4.0.

Experimental Results

Table 7.12 delineates the parameter configurations pertinent to the discussed model. To evaluate the predictive efficacy of this model, it is compared against five baseline methods across the four distinct seasons: winter, spring, summer, and autumn. The baseline methods include multi-layer perceptron (MLP), SVM, RNN, LSTM network, and WPD-LSTM model.

Seasonal fluctuations significantly influence the generation of PV power. Consequently, the discussed model, along with five baseline methods, is assessed across various seasons. In accordance with the seasonal classification in Australia, the test dataset is divided into four subsets: autumn (March–May), summer (December–February), spring (September–November), and winter (June–August).

Figures 7.19–7.22 illustrate the forecasting outcomes of the discussed method across different seasons. The predicted and original curves for each subset of PV power are displayed on the top, while the corresponding absolute error curves are presented on the bottom. Figures 7.19–7.22 reveal that the forecasting curves for all four seasons closely align with the original curves, particularly demonstrating optimal fit during the spring season. This demonstrates that the discussed model effectively captures the underlying trends in the data. Furthermore, it is observed that the absolute prediction errors are comparatively lower across all seasons.

Tables 7.13–7.16 present the results of the discussed model alongside the five baseline models in four seasons, spring, summer, autumn, and winter, which include WPD-LSTM, LSTM, GRU, RNN, and MLP, as referenced in Ref. [31]. The comparative analysis indicates that the discussed method surpasses all five baseline models in terms of the metrics MBE, MAPE, and RMSE across all seasons. For instance, the RMSE of the proposed model is 0.0692, which is markedly lower than the values of 1.0382, 1.035, 1.0581, 1.0861, and 0.2357 for LSTM, GRU, RNN, MLP, and WPD-LSTM, respectively. Similar trends are observed for the other two metrics, MAPE and MBE.

The predicted results for the test set reveal that the MBE metric for the proposed model varies between −0.0013 and −0.0074, with the absolute values of these extremes remaining within 1%. This suggests that the model exhibits robust

Table 7.12 Parameters of the discussed mode [26]/MDPI/CC BY 4.0.

Seasons	η	γ	n	μ	δ(%)	λ
Winter	0.01	0.4	51	0.98	9	1.0e−06
Autumn	0.01	0.4	51	0.98	9	1.0e−06
Summer	0.01	0.4	51	0.98	9	1.0e−06
Spring	0.01	0.4	51	0.98	9	1.0e−06

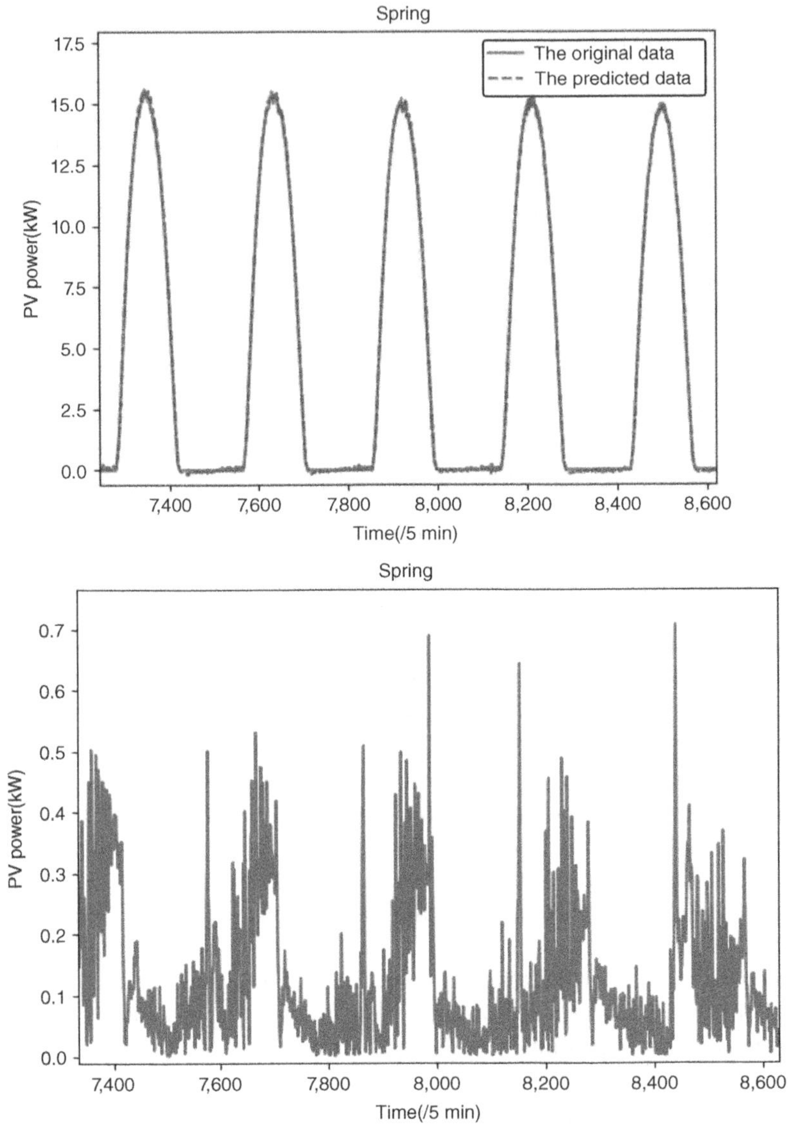

Figure 7.19 Prediction results of the discussed model in spring [26]/MDPI/CC BY 4.0.

fitting capabilities. In comparison to these baseline models, the discussed model exhibits a markedly lower absolute value of the MBE metric, indicating minimal prediction bias. Consistent conclusions are drawn for the MAPE and RMSE metrics when averaged across the entire test set.

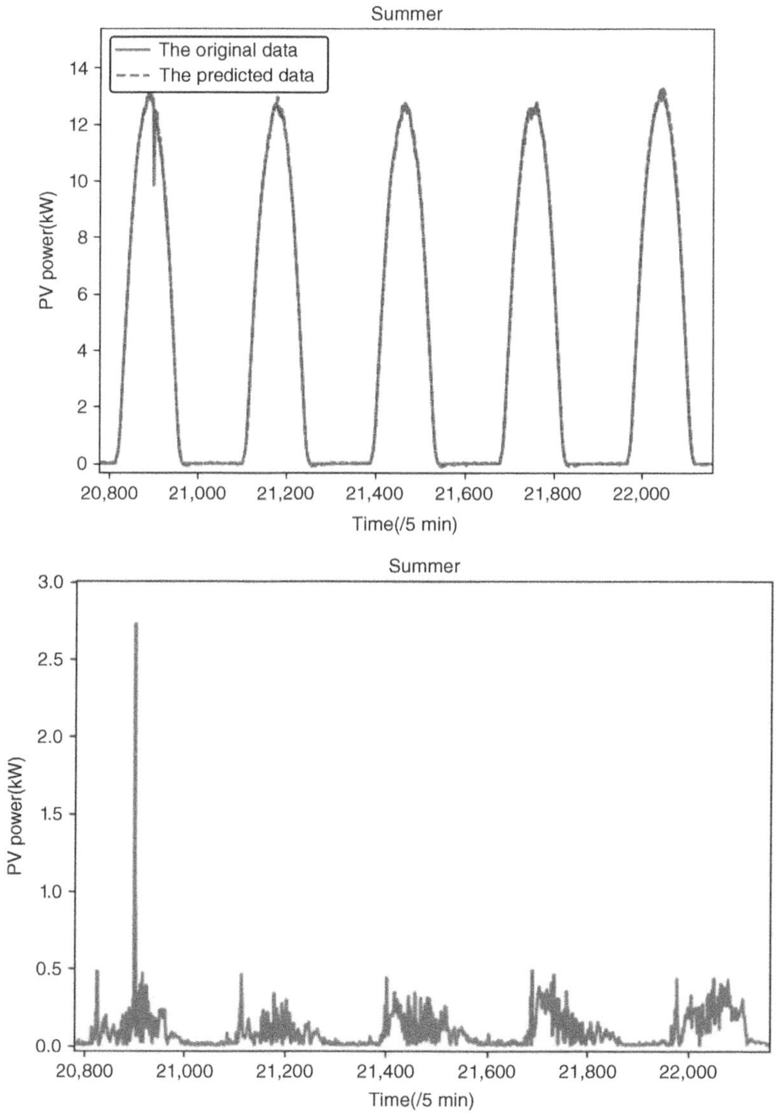

Figure 7.20 Forecasting results of the discussed model in summer [26]/MDPI/CC BY 4.0.

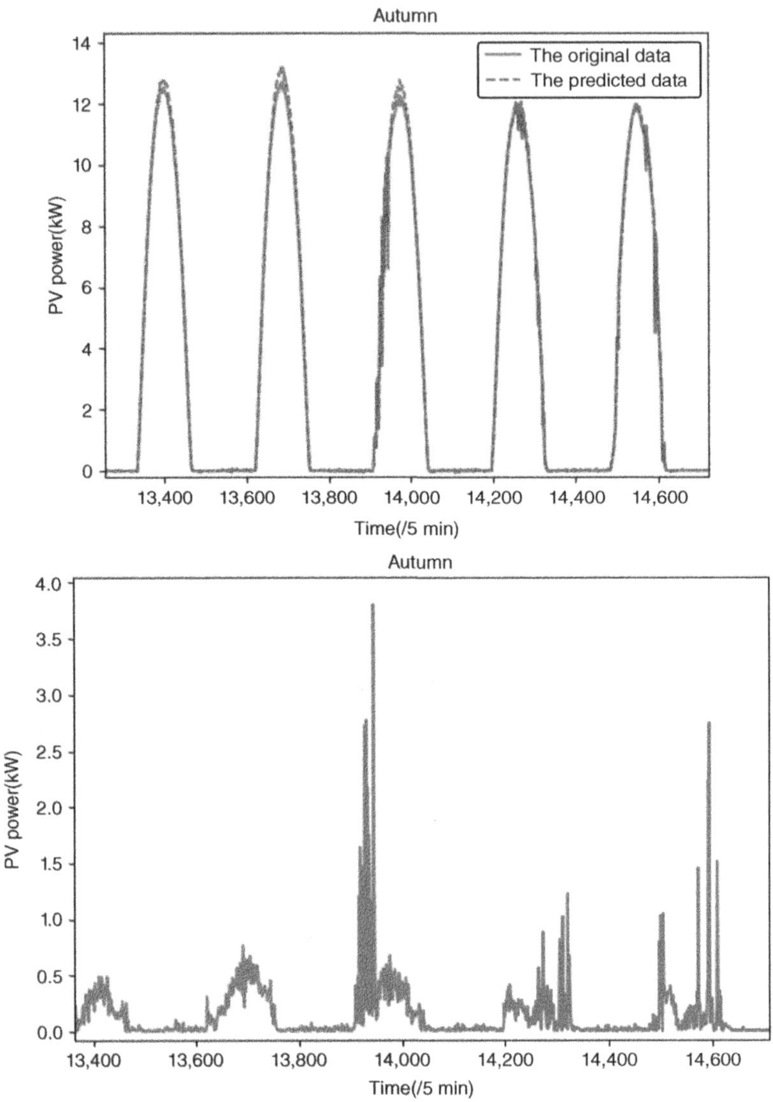

Figure 7.21 Forecasting results of the discussed model in autumn [26]/MDPI/CC BY 4.0.

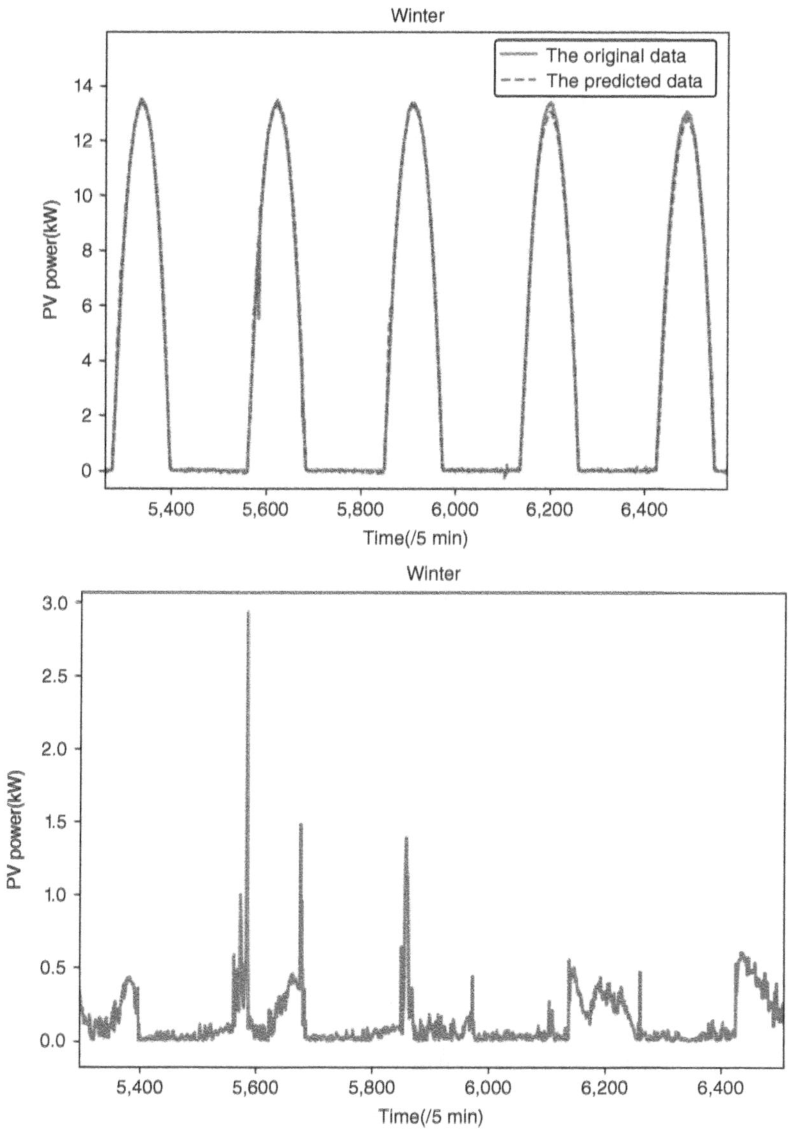

Figure 7.22 Forecasting results of the discussed model in winter [26]/MDPI/CC BY 4.0.

Table 7.13 Prediction results of distinct methods in spring. Adapted from [26].

Models	MBE	MAPE	RMSE
WPD-LSTM	0.0809	2.2660	0.2454
LSTM	−0.0767	5.1596	
RNN	−0.0667	6.1352	0.9340
MLP	0.3629	9.3335	1.0698
Our model	−0.0030	0.2816	0.0600

Table 7.14 Prediction results of different models in summer. Adapted from [26].

Models	MBE	MAPE	RMSE
WPD-LSTM	−0.0436	2.8885	0.2705
LSTM	−0.0792	11.1108	1.2504
GRU	0.2067	12.3369	1.2388
RNN	−0.2108	12.1321	1.2569
MLP	0.3535	12.4742	1.2630
Our model	−0.0074	0.5025	0.0819

Table 7.15 Prediction results of different models in autumn. Adapted from [26].

Models	MBE	MAPE	RMSE
WPD-LSTM	−0.0436	2.6219	0.2221
LSTM	−0.1268	5.6311	1.0710
GRU	0.1201	5.460	1.0748
RNN	−0.1718	7.7458	1.1022
MLP	0.0586	10.5393	1.0612
Our model	−0.0074	0.4524	0.0711

Table 7.16 Prediction results of different models in winter. Adapted from [26].

Models	MBE	MAPE	RMSE
WPD-LSTM	0.0396	1.8681	0.1526
LSTM	−0.0474	5.0221	0.8556
GRU	0.0600	5.6791	0.8471
RNN	−0.1254	6.4869	0.8810
MLP	0.0451	8.4689	0.9161
Our model	−0.0030	0.2816	0.0600

References

1 E.E. Elattar, J. Goulermas, and Q.H. Wu. Electric load forecasting based on locally weighted support vector regression. *IEEE Transactions on Systems, Man, and Cybernetics, Part C*, 40: 438–447, 2010.

2 L. Tang, X. Wang, X. Wang, C. Shao, S. Liu, and S. Tian. Long-term electricity consumption forecasting based on expert prediction and fuzzy Bayesian theory. *Energy*, 167: 1144–1154, 2019.

3 M.Q. Raza and A. Khosravi. A review on artificial intelligence based load demand forecasting techniques for smart grid and buildings. *Renewable and Sustainable Energy Reviews*, 50: 1352–1372, 2015.

4 A.O. Hoori, A. Al Kazzaz, R. Khimani, Y. Motai, and A.J. Aved. Electric load forecasting model using a multicolumn deep neural networks. *IEEE Transactions on Industrial Electronics*, 67(8): 6473–6482, 2019.

5 M. Imani. Electrical load-temperature CNN for residential load forecasting. *Energy*, 227: 120480, 2021.

6 M. Haris, M.N. Hasan, and S. Qin. Early and robust remaining useful life prediction of supercapacitors using BOHB optimized Deep Belief Network. *Applied Energy*, 286: 116541, 2021.

7 M. Alipour, J. Aghaei, M. Norouzi, T. Niknam, S. Hashemi, and M. Lehtonen. A novel electrical net-load forecasting model based on deep neural networks and wavelet transform integration. *Energy*, 205: 118106, 2020.

8 Z.A. Khan, T. Hussain, A. Ullah, S. Rho, M. Lee, and S.W. Baik. Towards efficient electricity forecasting in residential and commercial buildings: a novel hybrid CNN with a LSTM-AE based framework. *Sensors*, 20(5): 1399, 2020.

9 Z. Zhao, Y. Zhang, Y. Yang, and S. Yuan. Load forecasting via grey model-least squares support vector machine model and spatial-temporal distribution of electric consumption intensity. *Energy*, 255: 124468, 2022.

10 M. Ionescu, Gh. Păun, and T. Yokomori. Spiking neural P systems. *Fundamenta Informaticae*, 71: 279–308, 2006.

11 H. Peng, Z. Lv, B. Li, X. Luo, J. Wang, X. Song, T. Wang, M.J. Pérez-Jiménez, and A. Riscos-Núñez. Nonlinear spiking neural P systems. *International Journal of Neural Systems*, 30(10): 2050008, 2020.

12 L. Guo, J. Wang, H. Peng, and Y. Gao. Medium-long-term electricity load forecasting based on NSNP systems and attention mechanism. *Journal of Membrane Computing*, 6(1): 16–28, 2024.

13 W. Zhang, Q. Chen, J. Yan, S. Zhang, and J. Xu. A novel asynchronous deep reinforcement learning model with adaptive early forecasting method and reward incentive mechanism for short-term load forecasting. *Energy*, 236: 121492, 2021.

14 S. Li, P. Wang, and L. Goel. A novel wavelet-based ensemble method for short-term load forecasting with hybrid neural networks and feature selection. *IEEE Transactions on Power Systems*, 31(3): 1788–1798, 2015.

15 S. Li, L. Goel, and P. Wang. An ensemble approach for short-term load forecasting by extreme learning machine. *Applied Energy*, 170: 22–29, 2016.

16 J. Nowotarski, B. Liu, R. Weron, and T. Hong. Improving short term load forecast accuracy via combining sister forecasts. *Energy*, 98: 40–49, 2016.

17 M.Q. Raza, N. Mithulananthan, J. Li, and K.Y. Lee. Multivariate ensemble forecast framework for demand prediction of anomalous days. *IEEE Transactions on Sustainable Energy*, 11(1): 27–36, 2018.

18 G. Cervone, L. Clemente-Harding, S. Alessandrini, and L. Delle-Monache. Short-term photovoltaic power forecasting using artificial neural networks and an analog ensemble. *Renewable Energy*, 108: 274–286, 2017.

19 D.W. Van-der Meer, M. Shepero, A. Svensson, J. Widén, and J. Munkhammar. Probabilistic forecasting of electricity consumption, photovoltaic power generation and net demand of an individual building using gaussian processes. *Applied Energy*, 213: 195–207, 2018.

20 J. Antonanzas, N. Osorio, R. Escobar, R. Urraca, F.J. Martinez-de Pison, and F. Antonanzas-Torres. Review of photovoltaic power forecasting. *Solar Energy*, 136: 78–111, 2016.

21 M.Q. Raza and A. Khosravi. A review on artificial intelligence-based load demand forecasting techniques for smart grid and buildings. *Renewable and Sustainable Energy Reviews*, 50: 1352–1372, 2015.

22 R.A. de Marcos, A. Bello, and J. Reneses. Electricity price forecasting in the short-term hybridising fundamental and econometric modelling. *Electric Power Systems Research*, 167: 240–251, 2019.

23 S. Ben-Taieb, G. Bontempi, A.F. Atiya, and A. Sorjamaa. A review and comparison of strategies for multi-step ahead time series forecasting based on the nn5 forecasting competition. *Expert Systems with Applications*, 39: 7067–7083, 2012.

24 Q. Liu, L. Long, H. Peng, J. Wang, Q. Yang, X. Song, A. Riscos-Núñez, and M.J. Pérenz-Jiménez. Gated spiking neural P systems for time series forecasting. *IEEE Transactions on Neural Networks and Learning Systems*, 34(9): 6227–6236, 2023.

25 L. Long, R. Lugu, X. Xiong, Q. Liu, H. Peng, J. Wang, D. Orellana-Martín, and M.J. Pérez-Jiménez. Echo spiking neural P system. *Knowledge-Based Systems*, 253: 109568, 2022.

26 Y. Gao, J. Wang, L. Guo, and H. Peng. Short-term photovoltaic power prediction using nonlinear spiking neural P systems. *Sustainability*, 16(4): 1709, 2024.

27 Desert Knowledge Australia Centre. http://dkasolarcentre.com.au/historical-data/download.

28 S.L. Oh, E.Y.K. Ng, R.S. Tan, and U.R. Acharya. Automated diagnosis of arrhythmia using combination of CNN and LSTM techniques with variable length heart beats. *Computers in Biology and Medicine*, 102: 278–287, 2018.

29 K. Wang, X. Qi, and H. Liu. A comparison of day-ahead photovoltaic power forecasting models based on deep learning neural network. *Applied Energy*, 251: 113315, 2019.

30 X. Ren, F. Zhang, H. Zhu, and Y. Liu. Quad-kernel deep convolutional neural network for intra-hour photovoltaic power forecasting. *Applied Energy*, 323: 119682, 2022.

31 P. Li, K. Zhou, X. Lu, and S. Yang. A hybrid deep learning model for short-term PV power forecasting. *Applied Energy*, 259: 114216, 2020.

8

Medical Signal Processing

8.1 Introduction

As a widespread chronic neurological ailment, epilepsy manifests through unexpected erratic neuronal activity that disrupts normal brain operations temporarily, making it one of the planet's frequently diagnosed neural pathologies [1]. According to the World Health Organization, approximately 50 million people worldwide suffer from seizure disorders, which affect the normal life and physiological activities of people of all ages and may even be life-threatening. Therefore, the diagnosis and treatment of seizure is extremely important and of great clinical significance. Electroencephalography (EEG) is an effective method of monitoring brain activity, generated by neuronal discharges in the cerebral cortex. It has now become an effective clinical tool for seizure monitoring and treatment [2]. However, judging whether a patient suffers from a seizure by looking at the brainwave signals needs to be judged by a professional neurologist, which requires a lot of time and effort, and there are also some cases of inaccurate judgments, which all depend on the doctor's level of professionalism and proficiency in observing the brainwave signals. Moreover, doctors' observation methods are prone to some subjective errors and is quite a time-consuming and tedious process [3]. In today's era of continuous technological advancement, more and more industries have been replaced by machines, gradually changing from manual production to machine generation. The medical field is no exception, more and more artificial intelligence is applied to the medical field. The field of seizure detection has also seen the entry of AI. The addition of AI has greatly saved the time doctors spend judging and monitoring the brain waves of epilepsy patients.

EEG signal analysis has historically employed classical machine learning techniques [4]. Machine learning has become a hot topic in the research of seizure detection by brain waves in the last years [5–7]. These methods include

Spiking Neural P Systems for Time Series Analysis, First Edition. Jun Wang and Hong Peng.

k-means clustering, extreme gradient boost (XGBoost), support vector machine (SVM), and so on. Moreover, several signal decomposition tools have been introduced for epileptic seizure detection, for example, empirical mode decomposition (EMD), discrete wavelet transform (DWT), and local mean decomposition (LMD). Although machine learning methods have facilitated research in seizure detection, the limitations of them cannot be ignored, for example, the classifiers are manually designed and are not available for some of the objects.

As machine learning progresses, deep learning approaches have demonstrated notable efficacy in detecting epileptic conditions. These deep learning methods include recurrent neural networks (such as LSTM and GRU), graph convolutional networks (GCN) and convolutional neural networks (CNN). Despite the great progress in the problem of seizure detection using deep learning, there are still some limitations in the existing methods, for example, CNN-based methods are good at extracting local features, but there is no way to capture global dependencies. Recurrent neural networks such as LSTM can capture long-term dependencies in temporal data, but the performance aspect still needs to be improved.

This chapter will discuss a new seizure detection method, which utilizes multichannel LSTM-SNP model and uses DWT transform as the signal decomposition tool. In this chapter, the detection method is termed as DWT+LSTM-SNP [8].

8.2 Methodology

This part elaborates thoroughly on the proposed approach for detecting seizure episodes. Figure 8.1 shows the framework of the DWT+LSTM-SNP method [8].

The data is first preprocessed and the dataset is sliced into 1-second segments. The DWT technique is used to complete the 1-second segments, and the db4 wavelet is used for five-level decomposition. Then, the time domain features of each segment are extracted, and these features are put into a multichannel LSTM-SNP network. The multichannel LSTM-SNP is trained according to these feature and performs the seizure detection. The obtained results are tested and evaluated.

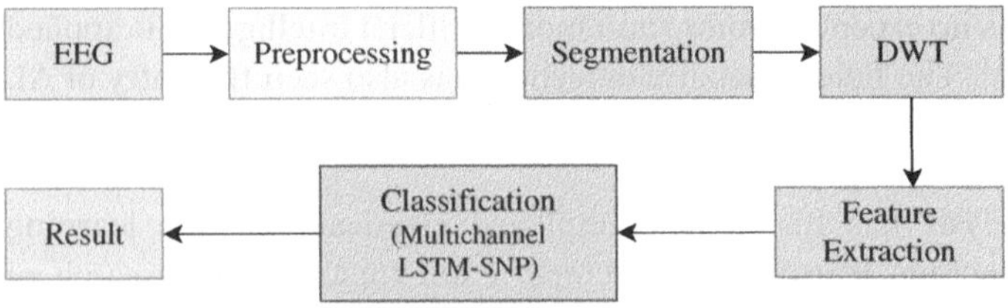

Figure 8.1 Framework of the DWT+LSTM-SNP method for seizure detection [8] / World Scientific Publishing Co Pte Ltd.

8.2.1 Dataset

The EEG dataset adopted is the CHB-MIT scalp EEG dataset collected from Boston Children's Hospital, which is publicly accessible on PhysioNet.org [9]. A total of 24 seizure patients are included in this dataset, of which 23 are patients with refractory seizure (17 females, aged 1.5–19 years, and 5 males, aged 3–22 years); the last case (chb24) has no clear gender or age recorded, so the brainwaves associated with this patient are not used; chb21 is a record of chb1 after 5 years, and is a record of chb1 from the same female patient [10]. Boston Children's Hospital evaluated the patient for potential conditions for surgical intervention several days after discontinuation of antiepileptic medication and monitored the patient for an extended period.

Raw scalp EEG recordings were made using a sampling rate of 256 Hz and 16-bit resolution. The electrical levels were placed and named according to the international 10-20 system electrode placement method [11]. Most patients have 23 channels of EEG signals, and a few patients have 24 or 26 channels of EEG signals. Each patient has 9–42 consecutively recorded EDF files of brainwave signals, and each EDF file records the patient's brainwave signals for up to 1 hour (3,600 seconds). There are individual cases with longer recorded EEG signals, e.g., chb04, chb07, and chb09 have EEG signals recorded for up to 4 hours. There are also individual patients who are recorded for a shorter period. However, due to hardware limitations, there is a discontinuity between each EDF file, and the time interval between each EDF is 10 seconds or even shorter or longer.

In view of the inconsistency in the number of channels recorded from different patients, 18 channels common to 23 patients are selected in this study, namely: FP1-F7, F7-T7, T7-P7, P7-O1, FP1-F3, F3-C3, C3-P3, P3-O1, FZ-CZ, CZ-PZ, FP2-F4, F4-C4, C4-P4, P4-O2, FP2-F8, F8-T8, T8-P8, and P8-O2. These electrodes cover the prefrontal, temporal, parietal, occipital, and central portions of the brain, as well as the region connecting the frontal and temporal lobes. Different electrodes are placed in different brain regions in order to perform seizure detection and identify abnormal discharge activity associated with seizures. The details of the CHB-MIT dataset is provided in Table 8.1.

8.2.2 Preprocessing

In order to improve the performance of the DWT+LSTM-SNP method on CHB-MIT dataset and to reduce the effect of noise on the EEG signals, the original EEG signals are denoised by band-pass filtering with range 0.5–50 Hz using the Butterworth zero-phase filtering algorithm. The band-pass filtering with range 0.5–50 Hz mainly removes or reduces the effect of noises from the power line, EMG, baseline drift, and other low and high-frequency noises, thus better highlighting the EEG activities related to seizure detection.

Table 8.1 The description of CHB-MIT dataset [8] / World Scientific Publishing Co Pte Ltd.

Number	Age	Sex	Number of seizures	Total duration of seizures (s)	Total time (hh:mm:ss)
chb23	6	F	7	424	26:33:30
chb22	9	F	3	204	31:00:11
chb21	13	F	4	199	32:49:49
chb20	6	F	8	294	27:36:06
chb19	19	F	3	236	29:55:46
chb18	18	F	6	317	35:38:05
chb17	12	F	3	293	21:00:24
chb16	7	F	10	84	19:00:00
chb15	16	M	20	1884	38:00:36
chb14	9	F	8	169	26:00:00
chb13	3	F	12	535	33:00:00
chb12	2	F	27	989	23:41:40
chb11	12	F	3	806	34:47:37
chb10	3	M	7	447	50:01:24
chb09	10	F	4	276	67:52:18
chb08	3.5	M	5	919	20:00:23
chb07	14.5	F	3	325	67:03:08
chb06	1.5	F	10	153	66:44:06
chb05	7	F	5	558	38:00:10
chb04	22	M	4	378	142:26:51
chb03	14	F	7	402	38:00:06
chb02	11	M	3	172	35:15:59
chb01	11	F	7	442	40:33:08
Total	—	—	169	10506	945:01:17

8.2.3 Discrete Wavelet Transform

Wavelet transform (WT) has been successfully applied to solve a variety of non-stationary signal problems, and it can more accurately reveal more details of the signal in the time and frequency domains [12]. The EEG signals are complex non-stationary signals with multiple frequency components, including signals from brain activities and physiological processes. Meanwhile, the features extracted in this study belong to the time domain, so it is appropriate and reasonable to use wavelet transform for EEG signal processing. As a widely adopted wavelet analysis method, the discrete wavelet transform (DWT) employs filter bank convolution to break down input signals into sets of orthogonal wavelets. As a result, EEG signals undergo decomposition via DWT to yield frequency-specific wavelet coefficients, followed by isolation of their unique features.

The DWT is given as follows:

$$W_f(a,b) = \frac{1}{\sqrt{|a|}} \int_{-\infty}^{+\infty} f(t)\bar{\psi}\left(\frac{t-b}{a}\right)dt \tag{8.1}$$

wherein ψ is the decomposed wavelet function; b and a denote the time translation and scale parameters, respectively.

After the DWT, the detail coefficients (dD) and approximation coefficients (dA) for each sub-band can be computed by

$$dD = \sum_{k=-\infty}^{+\infty} X[k]\varphi_l[2n-k] \tag{8.2}$$

$$dA = \sum_{k=-\infty}^{+\infty} X[k]\varphi_h[2n-k] \tag{8.3}$$

wherein X is the original brainwave signal; φ is the filter; h and l denote the high-pass and low-pass filters, respectively.

The purpose of multilevel decomposition by DWT is to obtain wavelet coefficients at different frequencies. The DWT decomposes the original input signal into an approximation coefficient and a detail coefficient at the first level, and then further decomposes the approximation coefficients into an approximation coefficient and detail coefficients, and sequentially decomposes the approximation coefficients [12]. The schematic diagram of multilevel decomposition performed by the DWT is shown in Figure 8.2.

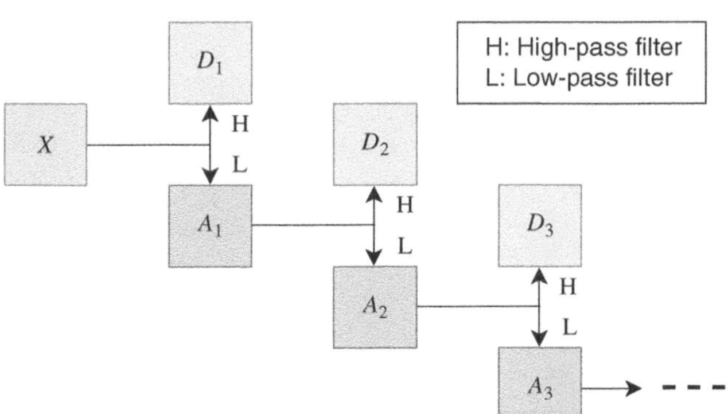

Figure 8.2 Schematic diagram of DWT multilevel decomposition [8] / World Scientific Publishing Co Pte Ltd.

Table 8.2 Frequency range corresponding to five-level decomposition [8] / World Scientific Publishing Co Pte Ltd.

Level	Modulus	Frequency range (Hz)
5	A5	0–4
5	D5	4–8
4	D4	8–16
3	D3	16–32
2	D2	32–64
1	D1	64–128

In the implementation, the "db4" wavelet is adopted to decompose the preprocessed EEG signals into five levels, and the wavelet coefficients in D1, D2, D3, D4, D5, and A5 obtained after decomposition correspond to different frequency ranges, shown in Table 8.2.

8.2.4 Feature Extraction

In this chapter, a five-level decomposition of the raw EEG signal is performed using "db4" wavelets. To find out the variability in different seizure states as well as to characterize the signals more comprehensively, 10 features are computed on the wavelet coefficients (D1, D2, D3, D4, and A4) of the five-level decomposition for a 1-second time window to capture the time-domain features of the signals, namely: minimum, maximum, mean, median, standard deviation, variance, mean absolute deviation, root mean square, skewness, and kurtosis. These features are defined by

$$\text{Maximum: } Max(X) \tag{8.4}$$

$$\text{Minimum: } Min(X) \tag{8.5}$$

$$\text{Median: } Median(X) \tag{8.6}$$

$$\text{Mean: } \bar{X} = \frac{\Sigma_{i=1}^{N} X(i)}{N} \tag{8.7}$$

$$\text{Standard: SD} = \sqrt{\frac{1}{N-1}\sum_{i=1}^{N}(X(i) - \bar{X})^2} \tag{8.8}$$

$$\text{Variance: V} = \frac{1}{N-1}\sum_{i=1}^{N}(X(i)-\bar{X})^2 \tag{8.9}$$

$$\text{Mean absolute deviation: MAD} = \frac{1}{N}\sum_{i=1}^{N}|X(i)-\bar{X}| \tag{8.10}$$

$$\text{Root mean square: RMS} = \sqrt{\frac{\sum_{i=1}^{N}(X(i))^2}{N}} \tag{8.11}$$

$$\text{Skewness: S} = \frac{\frac{1}{N}\sum_{i=1}^{N}(X(i)-\bar{X})^3}{\left(\sqrt{\frac{1}{N}\sum_{i=1}^{N}(X(i)-\bar{X})^2}\right)^3} \tag{8.12}$$

$$\text{Kurtosis: K} = \frac{\frac{1}{N}\sum_{i=1}^{N}(X(i)-\bar{X})^4}{\left(\sqrt{\frac{1}{N}\sum_{i=1}^{n}(X(i)-\bar{X})^2}\right)^2} \tag{8.13}$$

where X denotes an input signal, and N denotes a length of the input signal.

For each channel of each patient, the db4 wavelet technique is used to perform a five-level DWT decomposition to obtain five-level wavelet coefficients, and then 10 time-domain features are extracted for each wavelet coefficient, that is, each segment corresponds to $18 \times 5 \times 10 = 900$ features. The higher number of features demonstrates the variability in different seizure states in order to enable the model to learn more features and improve the performance of the model.

8.2.5 Multi-channel LSTM-SNP

As a specialized form of recurrent neural network, LSTM architecture was developed to address the vanishing gradient issue commonly encountered in standard RNN models. Its emergence has successfully solved the defects existing in the original RNNs, and it has become the most popular RNN variant at present [13].

As a neural computing model derived from NSNP systems, the LSTM-SNP framework exhibits LSTM-inspired recurrent characteristics suitable for sequential data analysis [14]. The features extracted in this study belong to the time domain, so this model can be used to train the features and detect seizure. By considering the variability of different feature values and the variability in different seizure states, a multichannel LSTM-SNP model is designed to deal with different feature values, so that the discussed model can cope with different seizure states and get better results for seizure detection. The architectural design of the multichannel LSTM-SNP approach is illustrated in Figure 8.3.

The LSTM-SNP demonstrates architectural similarities to standard LSTM networks in its recurrent framework, yet differs significantly in its mathematical representation (see Figure 8.4). To govern the recurrent framework, the LSTM-SNP system employs a triad of nonlinear gating units: generation, consumption, and reset gates [14]. Below are the specifications for the three nonlinear gate components.

1. Reset gate: Determine how many spikes in the previous network states are reset according to inputs from the current network and previous network states. The reset gate is defined by

$$r(\tau) = \rho(W_r \cdot x(\tau) + V_r \cdot v(\tau - 1) + b_r) \tag{8.14}$$

2. Consumption gate: Determine how many spikes in the previous network states are consumed based on inputs from the previous network and current

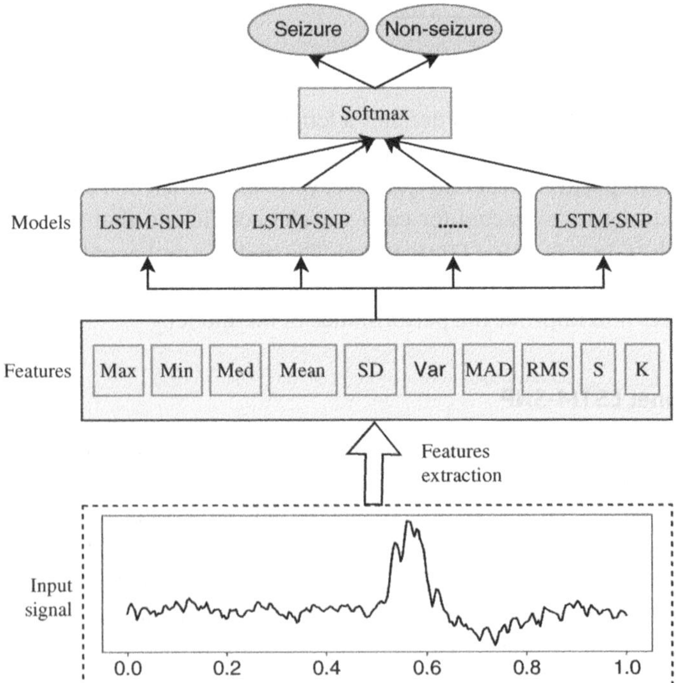

Figure 8.3 Structure of the multi-channel LSTM-SNP [8] / World Scientific Publishing Co Pte Ltd.

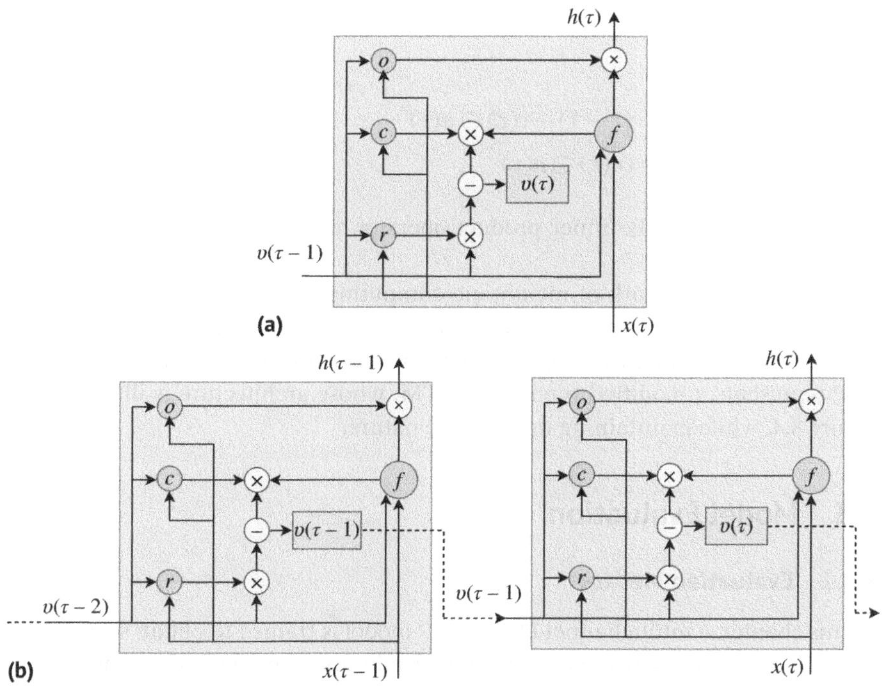

Figure 8.4 Structure of LSTM-SNP [8] / World Scientific Publishing Co Pte Ltd.

network states. The consumption gate is defined by

$$c(\tau) = \rho(W_c \cdot x(\tau) + V_c \cdot v(\tau - 1) + b_c) \tag{8.15}$$

3. Generation gate: Determine how many spikes are generated by the network according to inputs from the current network and previous network states. The generation gate is defined by

$$o(\tau) = \rho(W_o \cdot x(\tau) + V_o \cdot v(\tau - 1) + b_o) \tag{8.16}$$

The firing patterns produced by neuron σ are determined through

$$a(\tau) = f(W_a \cdot x(\tau) + V_a \cdot v(\tau - 1) + b_a) \tag{8.17}$$

where $v(\tau - 1)$ is the internal state of the neuron at time $\tau - 1$, and $x(\tau)$ is the input at time τ; in Eqs. (8.14)–(8.17), W_r, W_c, W_o, W_a,V_r, V_c, V_o, and V_a are the weight matrices, and b_r, b_c, b_o, and b_a are the bias vectors; $\rho(\cdot)$ is the nonlinear activation function, tan$h(\cdot)$ [14].

The state equation and input–output equation of the LSTM-SNP model is given by

$$\begin{cases} v(\tau) = r(\tau) \odot v(\tau - 1) - c(\tau) \odot a(\tau) \\ h(\tau) = o(\tau) \odot a(\tau) \end{cases} \tag{8.18}$$

wherein $\odot$ indicate the inner product operator, and $h(\tau)$ denotes the neuron's output.

Within the framework of membrane computing, the LSTM-SNP architecture employs a single nonlinear neuron with adjustable parameters, as illustrated in Figure 8.4 [14]. Viewed through the lens of recurrent neural networks, LSTM-SNP represents a modified version of LSTM, whose architecture is illustrated in Figure 8.4, while maintaining its trainable nature.

8.3 Model Evaluation

8.3.1 Evaluation Metrics

In this chapter, a multichannel LSTM-SNP model is trained to obtain seizure and non-seizure detection results. To evaluate the discussed method, three metrics are used in experiments. The three metrics are specificity, sensitivity and accuracy, which are given by

$$\text{Accuracy} = \frac{\text{TP} + \text{TN}}{\text{TP} + \text{FP} + \text{TN} + \text{FN}} \times 100\% \tag{8.19}$$

$$\text{Sensitivity} = \frac{\text{TP}}{\text{TP} + \text{FN}} \times 100\% \tag{8.20}$$

$$\text{Specificity} = \frac{\text{TN}}{\text{TN} + \text{FP}} \times 100\% \tag{8.21}$$

where FN denotes the number of seizure samples that are identified as non-seizure; TP denotes the number of seizure samples that are correctly identified; TN denotes the number of non-seizure samples that are correctly identified; FP denotes the number of non-seizure samples that are identified as seizure.

8.3.2 Dataset Partitioning

For computational convenience, the EDF file of each patient record is segmented into 1-second fragments, thus the 1-hour record is divided into 3,600 fragments. Then DWT transform is used for each fragment to extract 10 time domain features of wavelet coefficients, marking whether they are seizure or not.

Table 8.3 Division of the dataset [8] / World Scientific Publishing Co Pte Ltd.

	Seizure (hh:mm:ss)	Non-seizure (hh:mm:ss)	Total (hh:mm:ss)
Train	1:27:33	1:27:33	2:55:06
Test	2:55:06	942:06:11	945:01:17

Finally, the features are put into a multichannel LSTM-SNP model for training before seizure detection. In the experiments, up to 945 hours of the CHB-MIT dataset is adopted, but considering the imbalance in the ratio of seizure time to non-epileptic time as well as the variability of seizures, half of the total epileptic time is chosen as the epileptic versus non-epileptic time for the training dataset of up to 5,253 seconds (1:27:33). The seizure and nonseizure datasets are drawn from the total database and are randomly selected by the system to ensure the diversity of recorded signals. The dataset division is detailed in Table 8.3.

8.3.3 Experimental Results and Discussion

For every patient in the dataset, the results of the DWT+LSTM-SNP method are shown in Table 8.4. The last row of Table 8.4 demonstrates that the DWT+LSTM-SNP method has an average accuracy of 98.25%, an average sensitivity of 97.59%, and an average specificity of 98.22% in 23 patients.

For patient #12 the accuracy is lower mainly due to the fact that when patient #12 is recording the EDF file, the naming of the channels in the EDF file from 27 to 29 recording file is not consistent with the 18 common channels, leading to not recording the seizure signals in that part of the status epilepticus and also due to the fact that there is a large amount of external noise in the EEG signals of the patient, among other things. For patient #4 the results obtained by the DWT+LSTM-SNP method are lower than the others, mainly due to the fact that the signals are recorded over a longer period of time and the diversity of the signals is more complex, as well as having more noise in the signals, which result in not being able to filter out all of the noise when pre-processing the brainwave signals. The 100% accuracy for patient #6 is mainly due to the fact that the noise of patient #6 is removed by the Butterworth zero-phase filtering algorithm, and the wavelet coefficients have the least amount of noise.

The DWT+LSTM-SNP method is based on DWT and LSTM-SNP, and it is used in the public dataset CHB-MIT for seizure detection. Other seizure detection methods are also used for this dataset. Table 8.5 compares the DWT+LSTM-SNP method with other seizure detection methods on this dataset.

Table 8.4 Seizure detection results of the DWT+LSTM-SNP method [8] / World Scientific Publishing Co Pte Ltd.

Number	Accuracy (%)	Sensitivity (%)	Specificity (%)
chb01	99.02	95.65	100
chb02	97.81	94.32	99.21
chb03	100	100	99.1
chb04	92.88	92.39	95.33
chb05	97.88	95.65	96.89
chb06	100	100	100
chb07	94.32	98.92	98.11
chb08	96.69	96.93	99.11
chb09	97.41	96.85	97.78
chb10	100	98.78	98.32
chb11	98.93	96.35	96.52
chb12	94.27	94.31	96.93
chb13	98.56	96.34	95.63
chb14	99.21	100	100
chb15	98.31	98.78	97.96
chb16	100	99.65	98.52
chb17	99.52	95.7	96.01
chb18	99.25	99.51	98.26
chb19	100	99.38	98.34
chb20	99.36	99.19	99.19
chb21	100	99.78	99.78
chb22	99.76	98.89	99.53
chb23	96.65	97.12	98.62
Mean	98.25	97.59	98.22

Table 8.5 shows that the DWT+LSTM-SNP method outperforms most existing methods, including deep learning and machine learning models. This is mainly because some of these methods do not fully utilize the EEG signal channels to obtain effective features. However, in some evaluation metrics, it has not surpassed certain existing methods. For specificity and accuracy metrics, the DWT+LSTM-SNP did not surpass the Bi-GRU used by Ref. [16], but it far exceeded Bi-GRU in specificity metric, with a 3.7% increase. This is because it is important for seizure detection to be detected correctly, which can help doctors respond to epilepsy, and the DWT+LSTM-SNP method does

Table 8.5 Comparisons of the DWT+LSTM-SNP with other models [8] / World Scientific Publishing Co Pte Ltd.

Methods	Accuracy (%)	Sensitivity (%)	Specificity (%)
DWT+LSTM-SNP	98.25	97.59	98.22
[15]	97.18	97.10	97.77
[16]	98.49	93.89	98.49
[17]	97.83	98.24	97.39
[18]	96.67	97.72	95.62
[19]	96.05	96.05	96.61
[20]	97.09	96.81	97.26
[21]	95.51	95.14	94.6
[22]	91.19	84	99
[23]	—	93.61	91.85
[24]	97.49	97.34	97.51

not require postprocessing for seizure detection. In terms of sensitivity metric, the DWT+LSTM-SNP method did not surpass the models of Refs. [17, 18]. However, it is far superior to these two methods in terms of accuracy and specificity metrics. The specific response to non-epileptic detection levels is crucial for doctors to determine non-epileptic conditions, as the opposite is epileptic seizures, which also provides assistance for doctors. As the DWT+LSTM-SNP is a variant of LSTM-like models, more comparisons of LSTM models are also made in the experiments. For example, its sensitivity is 3.98% higher than the LMD+BiLSTM method proposed by Ref. [23]. Overall, the DWT+LSTM-SNP method is effective and stable for seizure detection on the CHB-MIT dataset, which validates its effectiveness and feasibility.

References

1 A.S. Zandi, M. Javidan, G.A. Dumont, and R. Tafreshi. Automated real-time epileptic seizure detection in scalp EEG recordings using an algorithm based on wavelet packet transform. *IEEE Transactions on Biomedical Engineering*, 57(7): 1639–1651, 2010.

2 K. Fu, J. Qu, Y. Chai, and T. Zou. Hilbert marginal spectrum analysis for automatic seizure detection in EEG signals. *Biomedical Signal Processing and Control*, 18:179–185, 2015.

3 M. Kaleem, A. Guergachi, and S. Krishnan. Patient-specific seizure detection in long-term EEG using wavelet decomposition. *Biomedical Signal Processing and Control*, 46:157–165, 2018.

4 A.F. Gill, S.A. Fatima, M. Usman Akram, S.G. Khawaja, and S.E. Awan. Analysis of EEG signals for detection of epileptic seizure using hybrid feature set. *Theory and Applications of Applied Electromagnetics: APPEIC 2014*, 344:49–57, 2015.

5 A. Emami, N. Kunii, T. Matsuo, T. Shinozaki, K. Kawai, and H. Takahashi. Autoencoding of long-term scalp electroencephalogram to detect epileptic seizure for diagnosis support system. *Computers in Biology and Medicine*, 110: 227–233, 2019.

6 N.S. Frolov, V.V. Grubov, V.A. Maksimenko, A. Luttjohann, V.V. Makarov, A.N. Pavlov, E. Sitnikova, A.N. Pisarchik, J. Kurths, and A.E. Hramov. Statistical properties and predictability of extreme epileptic events. *Scientific Reports*, 9(1): 7243, 2019.

7 R. San-Segundo, M. Gil-Martin, L.F. D'Haro-Enriquez, and J.M. Pardo. Classification of epileptic EEG recordings using signal transforms and convolutional neural networks. *Computers in Biology and Medicine*, 109: 148–158, 2019.

8 M. Wu, H. Peng, Z. Liu, and J. Wang. Seizure detection of EEG signals based on multi-channel long- and short-term memory-like spiking neural model. *International Journal of Neural Systems*, 34(10):106603, 2024.

9 B. Direito, J. Duarte, C. Teixeira, B. Schelter, M.L.V. Quyen, A. Schulze-Bonhage, F. Sales, and A. Dourado. Feature selection in high dimensional EEG features spaces for epileptic seizure prediction. *IFAC Proceedings Volumes*, 44:6206–6211, 2011.

10 K.M. Tsiouris, V.C. Pezoulas, M. Zervakis, S. Konitsiotis, D.D. Koutsouris, and D.I. Fotiadis. A long short-term memory deep learning network for the prediction of epileptic seizures using EEG signals. *Computers in Biology and Medicine*, 99: 24–37, 2018.

11 I. Guler and E.D. Ubeyli. Performance of epileptic single-channel scalp EEG classifications using single wavelet-based features. *Australasian Physical & Engineering Sciences in Medicine*, 40(1):57–67, 2017.

12 O. Faust, U.R. Acharya, H. Adeli, and A. Adeli. Wavelet-based EEG processing for computer-aided seizure detection and epilepsy diagnosis. *Seizure*, 26:56–64, 2015.

13 S. Hchreiter and J. Schmidhuber. Long short-term memory. *Neural Computation*, 9(8):1735–1780, 1997.

14 Q. Liu, L. Long, Q. Yang, H. Peng, J. Wang, and X. Luo. LSTM-SNP: a long short-term memory model inspired from spiking neural P systems. *Knowledge-Based Systems*, 235:107656, 2022.

15 H. Liu, Y. Gao, J. Zhang, and J. Zhang. Epilepsy EEG classification method based on supervised locality preserving canonical correlation analysis. *Mathematical Biosciences and Engineering*, 19(1):624–642, 2022.

16 Y. Zhang, S. Yao, R. Yang, X. Liu, W. Qiu, L. Han, W. Zhou, and W. Shang. Epileptic seizure detection based on bidirectional gated recurrent unit network. *IEEE Transactions on Neural Systems and Rehabilitation Engineering*, 30:135–145, 2022.

17 Y. Xiong, J. Li, D. Wu, F. Dong, J. Liu, L. Jiang, J. Cao, and Y. Xu. Seizure detection algorithm based on fusion of spatio-temporal network constructed with dispersion index. *Biomedical Signal Processing and Control*, 79:104155, 2023.

18 L. Jiang, J. He, H. Pan, D. Wu, T. Jiang, and J. Liu. Seizure detection algorithm based on improved functional brain network structure feature extraction. *Biomedical Signal Processing and Control*, 79:104053, 2023.

19 T. Dissanayake, T. Fernando, S. Denman, S. Sridharan, and C. Fookes. Geometric deep learning for subject independent epileptic seizure prediction using scalp EEG signals. *IEEE Journal of Biomedical and Health Informatics*, 26(2):527–538, 2021.

20 A. Zarei and B.M. Asl. Automatic seizure detection using orthogonal matching pursuit, discrete wavelet transform, and entropy based features of EEG signals. *IEEE Transactions on Neural Systems and Rehabilitation Engineering*, 131: 104250, 2021.

21 M. Shahbazi and H. Aghajan. A generalizable model for seizure prediction based on deep learning using CNN-LSTM architecture. *2018 IEEE Global Conference on Signal and Information Processing (GlobalSIP), IEEE*, pages 469–473, 2018.

22 W. Liang, H. Pei, Q. Cai, and Y. Wang. Scalp EEG epileptogenic zone recognition and localization based on long-term recurrent convolutional network. *Neurocomputing*, 396:569–576, 2020.

23 X. Hu, S. Yuan, F. Xu, Y. Leng, K. Yuan, and Q. Yuan. Scalp EEG classification using deep bi-LSTM network for seizure detection. *Computers in Biology and Medicine*, 124:103919, 2020.

24 C. Li, W. Zhou, G. Liu, Y. Zhang, M. Geng, Z. Liu, S. Wang, and W. Shang. Seizure onset detection using empirical mode decomposition and common spatial pattern. *IEEE Transactions on Neural Systems and Rehabilitation Engineering*, 29: 458–467, 2021.

Index

Spiking Neural P Systems for Time Series Analysis, First Edition. Jun Wang and Hong Peng.

Printed and bound by CPI Group (UK) Ltd, Croydon, CR0 4YY
17/09/2026
14964235-0001